The Britannica Guide to
Relativity and Quantum Mechanics

PHYSICS EXPLAINED

The Britannica Guide to Relativity and Quantum Mechanics

Edited by Erik Gregersen, Associate Editor, Science and Technology

IN ASSOCIATION WITH

Published in 2011 by Britannica Educational Publishing
(a trademark of Encyclopædia Britannica, Inc.)
in association with Rosen Educational Services, LLC
29 East 21st Street, New York, NY 10010.

Copyright © 2011 Encyclopædia Britannica, Inc. Britannica, Encyclopædia Britannica, and the Thistle logo are registered trademarks of Encyclopædia Britannica, Inc. All rights reserved.

Rosen Educational Services materials copyright © 2011 Rosen Educational Services, LLC. All rights reserved.

Distributed exclusively by Rosen Educational Services.
For a listing of additional Britannica Educational Publishing titles, call toll free (800) 237-9932.

First Edition

Britannica Educational Publishing
Michael I. Levy: Executive Editor
J.E. Luebering: Senior Manager
Marilyn L. Barton: Senior Coordinator, Production Control
Steven Bosco: Director, Editorial Technologies
Lisa S. Braucher: Senior Producer and Data Editor
Yvette Charboneau: Senior Copy Editor
Kathy Nakamura: Manager, Media Acquisition
Erik Gregersen: Associate Editor, Science and Technology

Rosen Educational Services
Nicholas Croce: Editor
Nelson Sá: Art Director
Cindy Reiman: Photography Manager
Matthew Cauli: Designer, Cover Design
Introduction by Erik Gregersen

Library of Congress Cataloging-in-Publication Data

The Britannica guide to relativity and quantum mechanics/edited by Erik Gregersen.
    p. cm. — (Physics explained)
"In association with Britannica Educational Publishing, Rosen Educational Services."
Includes bibliographical references and index.
ISBN 978-1-61530-330-4 (lib. bdg.)
 1. Relativity (Physics)—Popular works. 2. Quantum theory—Popular works.
I. Gregersen, Erik. II. Title: Guide to relativity and quantum mechanics.
III. Title: Relativity and quantum mechanics.
QC173.57.B75 2011
530.11—dc22

2010027855

Manufactured in the United States of America

On the cover, p. iii: Einstein's famous formula. *Shutterstock.com*

On page x: Composite image of warped space-time. *Victor de Schwanberg/Photo Researchers, Inc.*

On page xviii: A wormhole is solution of the field equations in Einstein's theory of general relativity that resembles a tunnel between two black holes. *Jean-Francois Podevin/Photo Researchers, Inc.*

On pages 1, 24, 51, 90, 112, 234, 237, 241: Matter from a star spiraling onto a black hole. *ESA, NASA, and Felix Mirabel (French Atomic Energy Commission and Institute for Astronomy and Space Physics/Conicet of Argentina)*

Contents

Introduction — x

Chapter 1: Relativity — 1

The Mechanical Universe — 1
Light and the Ether — 2
Special Relativity — 4
    Einstein's *Gedankenexperiments* — 4
    Starting Points and Postulates — 5
    Relativistic Space and Time — 6
    Relativistic Mass — 10
    Cosmic Speed Limit — 11
    $E = mc^2$ — 11
    The Twin Paradox — 11
    Four-Dimensional Space-Time — 12
    Experimental Evidence for Special Relativity — 22

Chapter 2: General Relativity — 24

Principle of Equivalence — 24
Curved Space-Time and Geometric Gravitation — 26
The Mathematics of General Relativity — 28
    Cosmological Solutions — 28
    Black Holes — 29
Experimental Evidence for General Relativity — 29
Unconfirmed Predictions of General Relativity — 31
    Gravitational Waves — 31
    Black Holes and Wormholes — 34
Applications of Relativistic Ideas — 35
    Elementary Particles — 35
    Particle Accelerators — 36
    Fission and Fusion: Bombs and Stellar Processes — 36
    The Global Positioning System — 37
    Cosmology — 37

Relativity, Quantum Theory,
and Unified Theories 46
Intellectual and Cultural Impact of
Relativity 47

Chapter 3: Quantum Mechanics: Concepts 51
Historical Basis of Quantum Theory 51
Early Developments 52
    Planck's Radiation Law 52
    Einstein and the Photoelectric
    Effect 53
    Bohr's Theory of the Atom 54
    Scattering of X-rays 58
    Broglie's Wave Hypothesis 59
Basic Concepts and Methods 60
Schrödinger's Wave Mechanics 61
Electron Spin and Antiparticles 64
Identical Particles and Multielectron
Atoms 69
Time-Dependent Schrödinger
Equation 74
Tunneling 76
Axiomatic Approach 78
Incompatible Observables 80
Heisenberg Uncertainty Principle 83
Quantum Electrodynamics 87

Chapter 4: Quantum Mechanics: Interpretation 90
The Electron: Wave or Particle? 90
Hidden Variables 92
Paradox of Einstein, Podolsky, and
Rosen 94
Measurement in Quantum Mechanics 98
Applications of Quantum Mechanics 101
    Decay of a Meson 101

Cesium Clock	104
A Quantum Voltage Standard	107
Bose-Einstein Condensate	109

Chapter 5: Biographies — 112

Carl David Anderson	112
Hans Bethe	113
David Bohm	118
Niels Bohr	120
Max Born	128
Satyendra Nath Bose	132
Louis-Victor, 7e duke de Broglie	132
Edward Uhler Condon	135
Clinton Joseph Davisson	137
P.A.M. Dirac	137
Sir Arthur Stanley Eddington	143
Albert Einstein	146
Enrico Fermi	163
Richard P. Feynman	169
Aleksandr Aleksandrovich Friedmann	173
George Gamow	174
Hans Geiger	176
Murray Gell-Mann	177
Walther Gerlach	179
Lester Halbert Germer	179
Samuel Abraham Goudsmit	180
Werner Heisenberg	182
Pascual Jordan	190
Brian D. Josephson	192
Max von Laue	194
Hendrik Antoon Lorentz	195
Ernst Mach	196
A.A. Michelson	198
Hermann Minkowski	201
Edward Williams Morley	202
Wolfgang Pauli	203
Max Planck	207

114

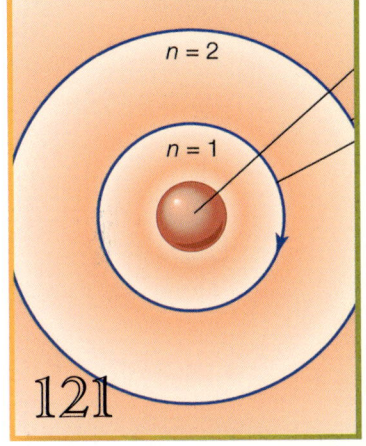

121

159

Henri Poincaré	215
Erwin Schrödinger	220
Karl Schwarzschild	223
Julian Seymour Schwinger	224
Arnold Sommerfeld	226
Otto Stern	227
Tomonaga Shin'Ichirō	229
George Eugene Uhlenbeck	230
Wilhelm Wien	231
Conclusion	232
Glossary	234
Bibliography	237
Index	241

Introduction

This volume deals with relativity and quantum mechanics. Both of these are quite new areas of physics. The beginning of relativity can be dated quite precisely, to the year 1905, when a clerk in the Swiss patent office published a paper "On the Electrodynamics of Moving Bodies." The beginnings of quantum mechanics can be dated to 1900 when the German physicist Max Planck explained the emission of light from a blackbody as the emission not of a continuous stream of particles or waves, but a stream of discrete packets of energy called quanta.

Relativity was driven by the need to explain light. The Scottish physicist James Clerk Maxwell had published four equations that explained electricity and magnetism. These equations described the speed of an electromagnetic wave. That speed was one with which scientists were already well acquainted. It was 299,000 km (186,000 miles) per second, the speed of light. Since light was an electromagnetic wave, it must be a wave in something, like waves in water or sound in air. As anyone who has ever looked up at the night sky knew, light crossed the vast emptiness of interstellar space from one star to another, which meant the vast emptiness was not empty at all. There was something there, something that had not been detected. This material, which came to be called the ether, had to be everywhere in the universe. Thomas Young said the ether pervaded "the substance of all material bodies as freely as wind passes through a grove of trees."

An American physicist named Albert Michelson devised an extremely clever experiment to detect the ether's effects. Light travelling in the same direction that Earth was moving through the solar system should be travelling at a speed that is the sum of two velocities: the velocity of Earth plus the velocity of light. Light traveling at a right angle to Earth's motion should just be traveling at the speed of light.

Michelson tried in 1881 to detect the difference in speed and failed. He tried again in 1887 with physicist Edward Morley an experiment that would detect differences much smaller than the 1881 experiment. There was no ether, and furthermore, in defiance of what everyone knew about physics, light traveled at exactly the same speed parallel or perpendicular to Earth's motion.

This result (or lack of a result) shattered physics. However, Einstein was undaunted by the end of classical physics. He took the invariance of the speed of light as one of his starting points for the theory of relativity. As another, he took that the laws of physics would look the same to all observers. From this foundation, Einstein developed the theory of special relativity.

When one first hears about the consequences of special relativity, they seem strange and hardly believable. Time runs more slowly in a moving object. Nothing can ever travel faster than light. However, these strange effects have been observed. Time dilation has been experimentally verified in many different ways. It has been tested by clocks on planes flying around the world and by particles entering Earth's atmosphere from outer space. The agreement between measurement and Einstein's theory has always been exact.

Of course, special relativity is "special" because it does not describe all motion. It did not describe any motion that is accelerated or decelerated. For example, any motion in a gravitational field experiences acceleration. It took Einstein 10 more years to solve the problem of acceleration, but he did with general relativity.

The results were as unusual as those of special relativity. Gravity was not a force but a bending of space-time, the very structure of the universe. Einstein himself was horrified by the fact that the equations of general relativity implied that the universe was expanding.

However, just as with special relativity, general relativity has been proven on many occasions. The first great test was looking for the deflection of starlight. In 1919, English expeditions went to West Africa and Brazil to observe a solar eclipse. General relativity passed the test. (This result was also seen as a triumph for science in that after the carnage of World War I, English scientists put aside national grudges to prove the theory of a German scientist.) Because each is very massive and move within the enormous gravitational field of the other, the effects of general relativity on the motion of the pulsars can be easily measured. General relativity has passed that test.

General relativity introduced new areas for astronomy to explore. Before his death in World War I, German astronomer Karl Schwarzschild found that the equations of relativity allowed an object in which mass was compressed into such a small space that the gravitational field would be so enormous that the velocity needed to break free of its gravitational influence would be larger than the speed of light, the cosmic speed limit. This object is called a black hole. (Although such a term is an obvious description, it was not so dubbed until 50 years later by American physicist John Wheeler.) Black holes are, of course, hard to observe directly, but there are many objects that seem to contain the requisite mass. One of these, Sagittarius A* (pronounced "A-star"), resides at the centre of the Milky Way Galaxy.

Despite Einstein's discomfort at the expanding universe, in the 1930s American astronomer Edwin Hubble had measured the distances to many galaxies and found that they were receding from the Milky Way at speeds proportional to their distances. This relation between speed and distance could only be explained by an expanding universe. Since the universe was expanding, this meant that early in its existence it was much much smaller and therefore hotter. This hot early universe is seen in the cosmic microwave background.

Relativity is a theory that applies to the large scale of the universe. The other subject of this book, quantum mechanics, is a theory of the extremely small. As with relativity, its results upend common sense notions of matter. Matter, in everyday experience, is solid, liquid, or gas. It is made up of atoms, which are usually drawn as miniature solar systems, with spheres of protons and neutrons in the center, orbited by moonlike electrons. This drawing does contain some truth but is as much metaphorical as actual. The protons and neutrons that make up the nucleus and the electrons around it sometimes have characteristics of both particles and waves.

Just like the surf pounding the beach or the light wave traveling through space, matter itself can be described as having a wave equation. This mathematical expression is called Schrdinger's equation, which contains a wave function that has values that depend on position. The square of this function is the probability of finding a particle at a position. This meant that on the subatomic scale, one could not say "the electron is here." The true statement is "the electron has this probability of being here. However, it may have a higher probability of being somewhere else." When this was applied to the hydrogen atom, it solved the mystery of why the electron only seemed to be in certain places within the atom. Any old function could not be a solution to Schrdinger's equation. Only certain functions (to be precise, products of Laguerre polynomials, which describe the part of the wave function that determines the distance from the nucleus, and spherical harmonics, which describe the part of the wave function that determines the angular part of the probability distribution) could actually solve the equation. These certain functions resulted in defined distances from the nucleus, or rather in distances where the square of the wave function was at a maximum.

When subatomic particles are considered as probabilities, they can do strange things, such as quantum tunneling. Suppose an electron requires some extra energy to get to the other side of some energy barrier. In ordinary mechanics, the electron has to have the extra energy or it is not going anywhere. However, with quantum mechanics, there is some probability that the electron could get through to the other side of the barrier without the extra energy. Sometimes this does happen. However it's more likely to happen if the amount of extra energy needed is not very much.

Another strange part of quantum mechanics was the uncertainty principle discovered by Werner Heisenberg. Suppose a physicist tries to measure the location of an electron. As the physicist measures the electron with greater and greater precision, the momentum of the electron is known with less and less precision. The converse is also true. Measurement of the momentum with greater precision leads to poorer knowledge of the position. In fact, the product of the uncertainties can never be less than a quantity called Planck's constant divided by 2 times pi. This was a somewhat disquieting result to some. There was a limit to what could be measured, and there was no way around the limit. Some physicists at the end of the 19th century said that their field would only consist of measuring what was already known to greater and greater precision. That was a pipe dream. Beyond a certain precision, one could go no further without throwing away other knowledge. There would always be a tradeoff.

There were quite a few physicists who were not happy with matter being constructed out of probabilities, with the universe as one giant casino. Einstein was chief among these and loudly asserted that "God does not play dice."

(Niels Bohr supposedly replied "Don't tell God what to do!") Einstein and other physicists sought "hidden" variables that underlay quantum mechanics and behaved in a more sensible way. However, no trace of the hidden variables have found, and the theories that postulate them are somewhat like the attempts of astronomers in the late Middle Ages to save the Earth-centred solar system by adding extremely complicated motions to it that would agree with the observations.

Both relativity and quantum mechanics arose in one of the great flowerings of science. In the early 20th century, scientists all over the world changed how humanity thought about how the universe began, how motion could be described, what matter was, and what the limits of physical knowledge were. The biographies of many of those who broke this new ground are in this volume. Much of today's physics, astronomy, and chemistry is following in the paths that these pioneers trailblazed.

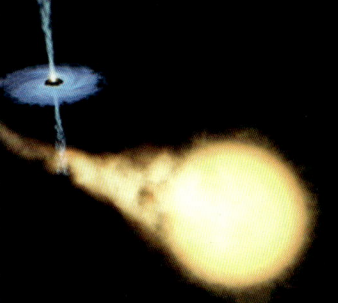

CHAPTER 1
RELATIVITY

With his theories of special relativity (1905) and general relativity (1916), German-born physicist Albert Einstein overthrew many assumptions underlying earlier physical theories, redefining in the process the fundamental concepts of space, time, matter, energy, and gravity. Along with quantum mechanics, relativity is central to modern physics. In particular, relativity provides the basis for understanding cosmic processes and the geometry of the universe itself.

THE MECHANICAL UNIVERSE

Relativity changed the scientific conception of the universe, which began in efforts to grasp the dynamic behaviour of matter. In Renaissance times, the great Italian physicist Galileo Galilei moved beyond Aristotle's philosophy to introduce the modern study of mechanics, which requires quantitative measurements of bodies moving in space and time. His work and that of others led to basic concepts, such as velocity, which is the distance a body covers in a given direction per unit time; acceleration, the rate of change of velocity; mass, the amount of material in a body; and force, a push or pull on a body.

The next major stride occurred in the late 17th century, when the British scientific genius Isaac Newton formulated his three famous laws of motion, the first and second of which are of special concern in relativity. Newton's first law, known as the law of inertia, states that a body that is not acted upon by external forces undergoes no acceleration—either remaining at rest or continuing to move in a straight line at constant speed. Newton's second law states that a force applied to a body changes its velocity by producing an acceleration that is proportional to the force and inversely proportional to the mass of the body. In constructing his system, Newton also defined space and time, taking both to be absolutes that are unaffected by anything external. Time, he wrote, "flows equably," while space "remains always similar and immovable."

Newton's laws proved valid in every application, as in calculating the behaviour of falling bodies, but they also provided the framework for his landmark law of gravity (the term, derived from the Latin *gravis*, or "heavy," had been in use since at least the 16th century). Beginning with the (perhaps mythical) observation of a falling apple and then considering the Moon as it orbits the Earth, Newton concluded that an invisible force acts between the Sun and its planets. He formulated a comparatively simple mathematical expression for the gravitational force; it states that every object in the universe attracts every other object with a force that operates through empty space and that varies with the masses of the objects and the distance between them.

The law of gravity was brilliantly successful in explaining the mechanism behind Kepler's laws of planetary motion, which the German astronomer Johannes Kepler had formulated at the beginning of the 17th century. Newton's mechanics and law of gravity, along with his assumptions about the nature of space and time, seemed wholly successful in explaining the dynamics of the universe, from motion on Earth to cosmic events.

LIGHT AND THE ETHER

However, this success at explaining natural phenomena came to be tested from an unexpected direction—the behaviour of light, whose intangible nature had puzzled philosophers and scientists for centuries. In 1873 the Scottish physicist James Clerk Maxwell showed that light is an electromagnetic wave with oscillating electrical and magnetic components. Maxwell's equations predicted that electromagnetic waves would travel through empty space at a speed of almost exactly 3×10^8 metres (186,000 miles) per second—i.e., according with the measured speed of light. Experiments soon confirmed the electromagnetic nature of light and established its speed as a fundamental parameter of the universe.

Maxwell's remarkable result answered long-standing questions about light, but it raised another fundamental issue: if light is a moving wave, what medium supports it? Ocean waves and sound waves consist of the progressive oscillatory motion of molecules of water and of atmospheric gases, respectively. But what is it that vibrates to make a moving light wave? Or to put it another way, how does the energy embodied in light travel from point to point?

For Maxwell and other scientists of the time, the answer was that light traveled in a hypothetical medium called the ether (aether). Supposedly, this medium permeated all space without impeding the motion of planets and stars; yet it had to be more rigid than steel so that light waves could move through it at high speed, in the same way that a taut guitar string supports fast mechanical vibrations. Despite this contradiction, the idea of the ether seemed essential—until a definitive experiment disproved it.

In 1887 the German-born American physicist A.A. Michelson and the American chemist Edward Morley made exquisitely precise measurements to determine how the Earth's motion through the ether affected the measured speed of light. In classical mechanics, the Earth's movement would add to or subtract from the measured speed of light waves, just as the speed of a ship would add to or subtract from the speed of ocean waves as measured from the ship. But the Michelson-Morley experiment had an unexpected outcome, for the measured speed of light remained the same regardless of the Earth's motion. This could only mean that the ether had no meaning and that the behaviour of light could not be explained by classical physics. The explanation emerged, instead, from Einstein's theory of special relativity.

SPECIAL RELATIVITY

"Special relativity" is limited to objects that are moving at constant speed in a straight line, which is called inertial motion. Beginning with the behaviour of light (and all other electromagnetic radiation), the theory of special relativity draws conclusions that are contrary to everyday experience but fully confirmed by experiments. Special relativity revealed that the speed of light is a limit that can be approached but not reached by any material object; it is the origin of the most famous equation in science, $E = mc^2$; and it has led to other tantalizing outcomes, such as the "twin paradox."

Einstein's *Gedankenexperiments*

Scientists such as Austrian physicist Ernst Mach and French mathematician Henri Poincaré had critiqued classical mechanics or contemplated the behaviour of light and the meaning of the ether before Einstein. Their efforts provided a background for Einstein's unique approach to understanding the universe, which he called in his native German a *Gedankenexperiment*, or "thought experiment."

Einstein described how at age 16 he watched himself in his mind's eye as he rode on a light wave and gazed at another light wave moving parallel to his. According to classical physics, Einstein should have seen the second light wave moving at a relative speed of zero. However, Einstein knew that Maxwell's electromagnetic equations absolutely require that light always move at 3×10^8 metres per second in a vacuum. Nothing in the theory allows a light wave to have a speed of zero. Another problem arose as well: if a fixed observer sees light as having a speed of 3×10^8 metres per second, whereas an observer moving at the speed of light sees light as having a speed of zero, it would mean that the laws of electromagnetism depend on the observer. But in classical mechanics the same laws apply for all observers, and Einstein saw no reason why the electromagnetic laws should not be equally universal. The constancy of the speed of light and the universality of the laws of physics for all observers are cornerstones of special relativity.

Starting Points and Postulates

In developing special relativity, Einstein began by accepting what experiment and his own thinking showed to be the true behaviour of light, even when this contradicted classical physics or the usual perceptions about the world.

The fact that the speed of light is the same for all observers is inexplicable in ordinary terms. If a passenger in a train moving at 100 km (60 miles) per hour shoots an arrow in the train's direction of motion at 200 km (120 miles) per hour, a trackside observer would measure the speed of the arrow as the sum of the two speeds, or 300 km (190 miles) per hour. In analogy, if the train moves at the speed of light and a passenger shines a laser in the same direction, then common sense indicates that a trackside observer should see the light moving at the sum of the two speeds, or twice the speed of light (6×10^8 metres [372,000 miles] per second).

While such a law of addition of velocities is valid in classical mechanics, the Michelson-Morley experiment showed that light does not obey this law. This contradicts common sense; it implies, for instance, that both a train moving at the speed of light and a light beam emitted from the train arrive at a point farther along the track at the same instant.

Nevertheless, Einstein made the constancy of the speed of light for all observers a postulate of his new theory. As a second postulate, he required that the laws of physics have the same form for all observers. Then Einstein extended his postulates to their logical conclusions to form special relativity.

Relativistic Space and Time

Since the time of Galileo it has been realized that there exists a class of so-called inertial frames of reference—i.e., in a state of uniform motion with respect to one another such that one cannot, by purely mechanical experiments, distinguish one from the other. It follows that the laws of mechanics must take the same form in every inertial frame of reference. To the accuracy of present-day technology, the class of inertial frames may be regarded as those that are neither accelerating nor rotating with respect to the distant galaxies. To specify the motion of a body relative to a frame of reference, one gives its position x as a function of a time coordinate t (x is called the position vector and has the components x, y, and z).

Newton's first law of motion (which remains true in special relativity) states that a body acted upon by no external forces will continue to move in a state of uniform motion relative to an inertial frame. It follows from this that the transformation between the coordinates (t, x) and (t', x') of two inertial frames with relative velocity u must be related by a linear transformation. Before Einstein's special theory of relativity was published in 1905, it was usually assumed that the time coordinates measured in all inertial frames were identical and equal to an "absolute time." Thus,

$$t = t'. \tag{1}$$

The position coordinates x and x' were then assumed to be related by

$$x' = x - ut. \tag{2}$$

The two formulas (1) and (2) are called a Galilean transformation. The laws of nonrelativistic mechanics take the same form in all frames related by Galilean transformations. This is the restricted, or Galilean, principle of relativity.

In order to make the speed of light constant, Einstein replaced absolute space and time with new definitions that depend on the state of motion of an observer. Einstein explained his approach by considering two observers and a train. One observer stands alongside a straight track; the other rides a train moving at constant speed along the track. Each views the world relative to his own surroundings. The fixed observer measures distance from a mark inscribed on the track and measures time with his watch; the train passenger measures distance from a mark inscribed on his railroad car and measures time with his own watch.

If time flows the same for both observers, as Newton believed, then the two frames of reference are reconciled by the relation: $x' = x - vt$. Here x is the distance to some specific event that happens along the track, as measured by the fixed observer; x' is the distance to the same event as measured by the moving observer; v is the speed of the train—that is, the speed of one observer relative to the other; and t is the time at which the event happens, the same for both observers.

For example, suppose the train moves at 40 km (25 miles) per hour. One hour after it sets out, a tree 60 km (37 miles) from the train's starting point is struck by lightning. The fixed observer measures x as 60 km and t as one hour. The moving observer also measures t as one hour, and so, according to Newton's equation, he measures x' as 20 km (12 miles).

This analysis seems obvious, but Einstein saw a subtlety hidden in its underlying assumptions—in particular, the issue of simultaneity. The two people do not actually observe the lightning strike at the same time. Even at the speed of light, the image of the strike takes time to reach each observer, and, since each is at a different distance from the event, the travel times differ. Taking this insight further, suppose lightning strikes two trees, one 60 km ahead of the fixed observer and the other 60 km behind, exactly as the moving observer passes the fixed observer. Each image travels the same distance to the fixed observer, and so he certainly sees the events simultaneously. The motion of the moving observer brings him closer to one event than the other, however, and he thus sees the events at different times.

Einstein concluded that simultaneity is relative; events that are simultaneous for one observer may not be for another. This led him to the counterintuitive idea that time flows differently according to the state of motion and to the conclusion that distance is also relative. In the example, the train passenger and the fixed observer can each stretch a tape measure from back to front of a railroad car to find its length. The two ends of the tape must be placed in position at the same instant—that is, simultaneously—to obtain a true value. However, because the meaning of simultaneous is different for the two observers, they measure different lengths.

This reasoning led Einstein to give up the Galilean transformations (1) and (2) and replace them with new equations for time and space, called the Lorentz transformations, after the Dutch physicist Hendrik Lorentz, who first proposed them. They are:

$$x' = \frac{x - vt}{\sqrt{1 - \frac{v^2}{c^2}}} \text{ and } t' = \frac{t - \frac{vx}{c^2}}{\sqrt{1 - \frac{v^2}{c^2}}},$$

where t' is time as measured by the moving observer and c is the speed of light.

From these equations, Einstein derived a new relationship that replaces the classical law of addition of velocities,

$$u' = \frac{u + v}{1 + \frac{uv}{c^2}},$$

where u and u' are the speed of any moving object as seen by each observer and v is again the speed of one observer relative to the other. This relation guarantees Einstein's first postulate (that the speed of light is constant for all observers). In the case of the flashlight beam projected from a train moving at the speed of light, an observer on the train measures the speed of the beam as c. According to the equation above, so does the trackside observer, instead of the value $2c$ that classical physics predicts.

To make the speed of light constant, the theory requires that space and time change in a moving body, according to its speed, as seen by an outside observer. The body becomes shorter along its direction of motion; that is, its length contracts. Time intervals become longer, meaning that time runs more slowly in a moving body; that is, time dilates. In the train example, the person next to the track measures a shorter length for the train and a longer time interval for clocks on the train than does the train passenger. The relations describing these changes are

$$L = L_0 \sqrt{1 - \frac{v^2}{c^2}} \text{ and } T = \frac{T_0}{\sqrt{1 - \frac{v^2}{c^2}}},$$

where L_o and T_o, called proper length and proper time, respectively, are the values measured by an observer on the moving body, and L and T are the corresponding quantities as measured by a fixed observer.

The relativistic effects become large at speeds near that of light, although it is worth noting again that they appear only when an observer looks at a moving body. He never sees changes in space or time within his own reference frame (whether on a train or spacecraft), even at the speed of light. These effects do not appear in ordinary life, because the factor v^2/c^2 is minute at even the highest speeds attained by humans, so that Einstein's equations become virtually the same as the classical ones.

Relativistic Mass

To derive further results, Einstein combined his redefinitions of time and space with two powerful physical principles: conservation of energy and conservation of mass, which state that the total amount of each remains constant in a closed system. Einstein's second postulate ensured that these laws remained valid for all observers in the new theory, and he used them to derive the relativistic meanings of mass and energy.

Cosmic Speed Limit

One result is that the mass of a body increases with its speed. An observer on a moving body, such as a spacecraft, measures its so-called rest mass m_0, while a fixed observer measures its mass m as

$$m = \frac{m_0}{\sqrt{1 - \frac{v^2}{c^2}}},$$

which is greater than m_0. In fact, as the spacecraft's speed approaches that of light, the mass m approaches infinity. However, as the object's mass increases, so does the energy required to keep accelerating it; thus, it would take infinite energy to accelerate a material body to the speed of light. For this reason, no material object can reach the speed of light, which is the speed limit for the universe. (Light itself can attain this speed because the rest mass of a photon, the quantum particle of light, is zero.)

$E = mc^2$

Einstein's treatment of mass showed that the increased relativistic mass comes from the energy of motion of the body—that is, its kinetic energy E—divided by c^2. This is the origin of the famous equation $E = mc^2$, which expresses the fact that mass and energy are the same physical entity and can be changed into each other.

The Twin Paradox

The counterintuitive nature of Einstein's ideas makes them difficult to absorb and gives rise to situations that seem unfathomable. One well-known case is the twin paradox, a seeming anomaly in how special relativity describes time.

Suppose that one of two identical twin sisters flies off into space at nearly the speed of light. According to relativity, time runs more slowly on her spacecraft than on Earth; therefore, when she returns to Earth, she will be younger than her Earth-bound sister. But in relativity, what one observer sees as happening to a second one, the second one sees as happening to the first one. To the space-going sister, time moves more slowly on Earth than in her spacecraft; when she returns, her Earth-bound sister is the one who is younger. How can the space-going twin be both younger and older than her Earth-bound sister?

The answer is that the paradox is only apparent, for the situation is not appropriately treated by special relativity. To return to Earth, the spacecraft must change direction, which violates the condition of steady straight-line motion central to special relativity. A full treatment requires general relativity, which shows that there would be an asymmetrical change in time between the two sisters. Thus, the "paradox" does not cast doubt on how special relativity describes time, which has been confirmed by numerous experiments.

Four-Dimensional Space-Time

Special relativity is less definite than classical physics in that both the distance D and time interval T between two events depend on the observer. Einstein noted, however, that a particular combination of D and T, the quantity $D^2 - c^2T^2$, has the same value for all observers. The term cT in this invariant quantity elevates time to a kind of mathematical parity with space. Noting this, the German mathematical physicist Hermann Minkowski showed that the universe resembles a four-dimensional structure with coordinates x, y, z, and ct representing length, width, height, and time, respectively. Hence, the universe can be described as a four-dimensional space-time continuum, a central concept in general relativity.

Minkoswki noted that the motion of a particle may be regarded as forming a curve, called a world line, made up of points, called events, in this space-time. It is frequently useful to represent physical processes by space-time diagrams in which time runs vertically and the spatial coordinates run horizontally. Of course, since space-time is four-dimensional, at least one of the spatial dimensions in the diagram must be suppressed.

Newton's first law can be interpreted in four-dimensional space as the statement that the world lines of particles suffering no external forces are straight lines in space-time. Linear transformations take straight lines to straight lines, and Lorentz transformations have the additional property that they leave invariant the invariant interval τ through two events $(t_1, \boldsymbol{x}_1)$ and $(t_2, \boldsymbol{x}_2)$ given by

$$\tau^2 = (t_1 - t_2)^2 - \frac{(\boldsymbol{x}_1 - \boldsymbol{x}_2)^2}{c^2}. \tag{3}$$

If the right-hand side of equation (3) is zero, the two events may be joined by a light ray and are said to be on each other's light cones because the light cone of any event (t,x) in space-time is the set of points reachable from it by light rays. Thus the set of all events $(t_2, \boldsymbol{x}_2)$ satisfying equation (3) with zero on the right-hand side is the light cone of the event $(t_1, \boldsymbol{x}_1)$. Because Lorentz transformations leave invariant the space-time interval (3), all inertial observers agree on what the light cones are. In space-time diagrams it is customary to adopt a scaling of the time coordinate such that the light cones have a half angle of 45°.

If the right-hand side of equation (3) is strictly positive, in which case one says that the two events are timelike

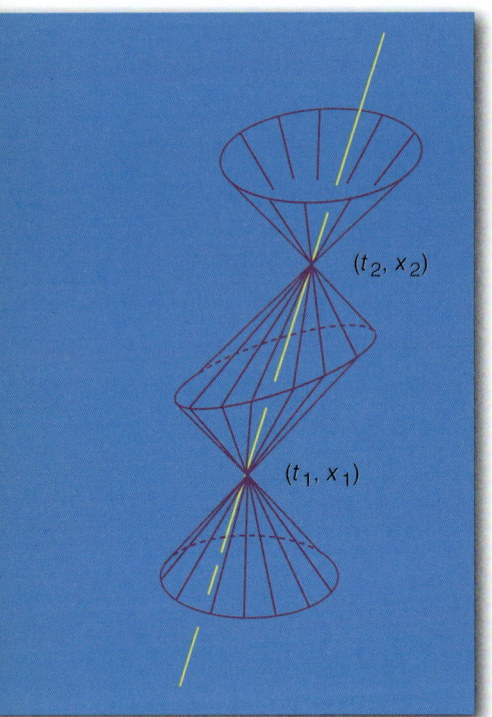

The world line of a particle traveling with speed less than that of light. Copyright Encyclopædia Britannica; rendering for this edition by Rosen Educational Services

separated, or have a timelike interval, then one can find an inertial frame with respect to which the two events have the same spatial position. The straight world line joining the two events corresponds to the time axis of this inertial frame of reference. The quantity τ is equal to the difference in time between the two events in this inertial frame and is called the proper time between the two events. The proper time would be measured by any clock moving along the straight world line between the two events.

An accelerating body will have a curved world line that may be specified by giving its coordinates t and x as a function of the proper time τ along the world line. The laws of either may be phrased in terms of the more familiar velocity $v = dbix/dt$ and acceleration $a = d^2x/dt^2$r in terms of the 4-velocity $(dt/d\tau, dbix/d\tau)$ and 4-acceleration $(d^2t/d\tau^2, dbix/d\tau^2)$. Just as an ordinary vector like v has three components, $v_x, v_y,$ and v_z, a 4-vector has four components. Geometrically the 4-velocity and 4-acceleration correspond, respectively, to the tangent vector and the curvature vector of the world line. If the particle moves slower than light, the tangent, or velocity, vector at each event on the world line points inside the light cone of that event, and the acceleration, or curvature, vector points outside the light cone. If the particle moves with the speed of light, then the tangent vector lies on the light cone at each event on the world line. The proper time τ along a world line moving with a speed less than light is not an independent quantity from *t* and ***x***: it satisfies

$$\left(\frac{dt}{d\tau}\right)^2 - \frac{1}{c^2}\left(\frac{d\mathbf{x}}{d\tau}\right)^2 = 1. \qquad (4)$$

For a particle moving with exactly the speed of light, one cannot define a proper time τ. One can, however, define a so-called affine parameter that satisfies equation (4) with zero on the right-hand side. For the time being this discussion will be restricted to particles moving with speeds less than light.

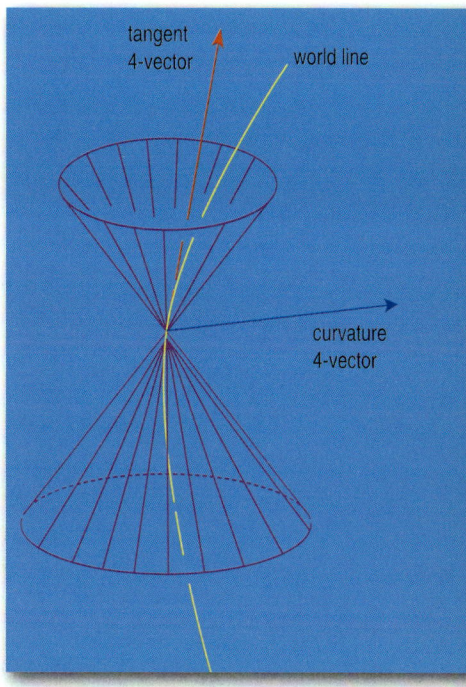

The world line of an accelerating body moving slower than the speed of light; the tangent vector corresponds to the body's 4-velocity and the curvature vector to its 4-acceleration. Copyright Encyclopædia Britannica; rendering for this edition by Rosen Educational Services

Equation (4) does not fix the sign of τ relative to that of *t*. It is usual to resolve this ambiguity by demanding that the proper time τ increase as the time *t* increases. This requirement is invariant under Lorentz transformations of the form of equations (1) and (2). The tangent vector then points inside the future light cone and is said to be future-directed and timelike. One may if one wishes attach an arrow to the world line to indicate this fact. One says that the particle moves forward in time. It was pointed out by the Swiss physicist Ernest C.G.

Stückelberg de Breidenbach and by the American physicist Richard Feynman that a meaning can be attached to world lines moving backward in time—i.e., for those for which ordinary time t decreases as proper time τ increases. Since the energy E of a particle is $mc^2 dt/d\tau$, such world lines correspond to the motion of particles with negative energy. It is possible to interpret these world lines in terms of antiparticles, as will be seen when particles moving in a background electromagnetic field are considered.

The fundamental laws of motion for a body of mass m in relativistic mechanics are

$$m\frac{d^2 t}{d\tau^2} = f^0 \qquad (5)$$

and

$$m\frac{d^2 x}{d\tau^2} = f, \qquad (6)$$

where m is the constant so-called rest mass of the body and the quantities (f^0, f) are the components of the force 4-vector. Equations (5) and (6), which relate the curvature of the world line to the applied forces, are the same in all inertial frames related by Lorentz transformations. The quantities $(mdt/d\tau, mdbix/d\tau)$ make up the 4-momentum of the particle. According to Minkowski's reformulation of special relativity, a Lorentz transformation may be thought of as a generalized rotation of points of Minkowski space-time into themselves. It induces an identical rotation on the 4-acceleration and force 4-vectors. To say that both of these 4-vectors experience the same generalized rotation or Lorentz transformation is simply to say that the fundamental laws of motion (5) and (6) are the same in all inertial frames related by

Lorentz transformations. Minkowski's geometric ideas provided a powerful tool for checking the mathematical consistency of special relativity and for calculating its experimental consequences. They also have a natural generalization in the general theory of relativity, which incorporates the effects of gravity.

The law of motion (6) may also be expressed as:

$$\frac{d}{dt}\left(\frac{mv}{\sqrt{1-v^2/c^2}}\right) = F, \qquad (7)$$

where $F = f(1 - v^2/c^2)$. Equation (7) is of the same form as Newton's second law of motion, which states that the rate of change of momentum equals the applied force. F is the Newtonian force, but the Newtonian relation between momentum p and velocity v in which $p = mv$ is modified to become

$$p = \frac{mv}{\sqrt{1-v^2/c^2}}. \qquad (8)$$

Consider a relativistic particle with positive energy and electric charge q moving in an electric field E and magnetic field B; it will experience an electromagnetic, or Lorentz, force given by $F = qE + qv \times B$. If $t(\tau)$ and $x(\tau)$ are the time and space coordinates of the particle, it follows from equations (5) and (6), with $f^0 = (qE \cdot v)dt/d\tau$ and $f = q(E + v \times B)dt/d\tau$, that $-t(-\tau)$ and $-x(-\tau)$ are the coordinates of a particle with positive energy and the opposite electric charge $-q$ moving in the same electric and magnetic field. A particle of the opposite charge but with the same rest mass as the original particle is called the original particle's antiparticle. It is in this sense that Feynman and Stückelberg spoke of antiparticles as particles moving backward in time. This idea is a consequence of special relativity alone. It really comes into its own, however, when one considers relativistic quantum mechanics.

Just as in nonrelativistic mechanics, the rate of work done when the point of application of a force $\boldsymbol{F}$ is moved with velocity $\boldsymbol{v}$ equals $\boldsymbol{F} \cdot \boldsymbol{v}$ when measured with respect to the time coordinate t. This work goes into increasing the energy E of the particle. Taking the dot product of equation (7) with $\boldsymbol{v}$ gives

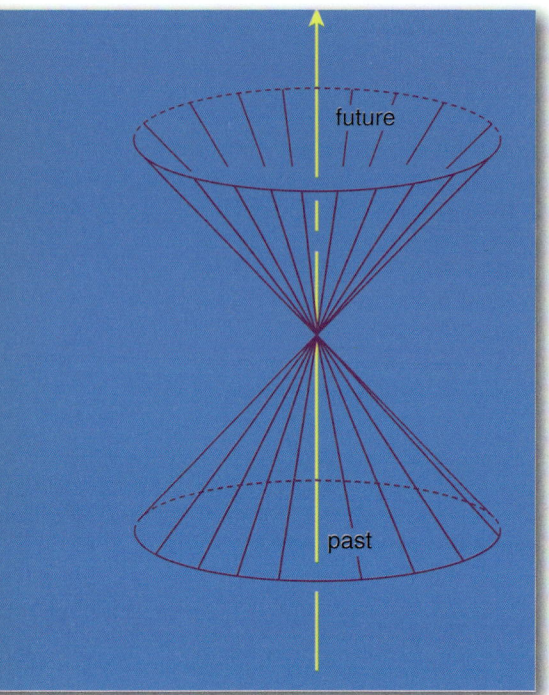

The world line of a particle moving forward in time. Copyright Encyclopædia Britannica; rendering for this edition by Rosen Educational Services

$$\frac{dE}{dt} = \boldsymbol{F} \cdot \boldsymbol{v},$$

where (9)

$$E = \frac{mc^2}{\sqrt{1 - v^2/c^2}} = mc^2 \frac{dt}{d\tau}.$$

The reader should note that the 4-momentum is just $(E/c^2, \boldsymbol{p})$. It was once fairly common to encounter the use of a "velocity-dependent mass" equal to E/c^2. However, experience has shown that its introduction serves no useful purpose and may lead to confusion, and it is not used in this article. The invariant quantity is the rest mass m. For that reason it has not been thought necessary to add a subscript or superscript to m to emphasize that it is the rest mass rather than a velocity-dependent quantity. When subscripts are attached to a mass, they indicate the particular particle of which it is the rest mass.

If the applied force F is perpendicular to the velocity v, it follows from equation (9) that the energy E, or, equivalently, the velocity squared v^2, will be constant, just as in Newtonian mechanics. This will be true, for example, for a particle moving in a purely magnetic field with no electric field present. It then follows from equation (7) that the shape of the orbits of the particle are the same according to the classical and the relativistic equations. However, the rate at which the orbits are traversed differs according to the two theories. If w is the speed according to the nonrelativistic theory and v that according to special relativity, then $w = v\sqrt{(1 - v^2/c^2)}$.

For velocities that are small compared with that of light,

$$E \approx mc^2 + \frac{1}{2}mv^2.$$

The first term, mc^2, which remains even when the particle is at rest, is called the rest mass energy. For a single particle, its inclusion in the expression for energy might seem to be a matter of convention: it appears as an arbitrary constant of integration. However, for systems of particles that undergo collisions, its inclusion is essential.

Both theory and experiment agree that, in a process in which particles of rest masses $m_1, m_2, \ldots m_n$ collide or decay or transmute one into another, both the total energy $E_1 + E_2 + \ldots + E_n$ and the total momentum $p_1 + p_2 + \ldots + p_n$ are the same before and after the process, even though the number of particles may not be the same before and after. This corresponds to conservation of the total 4-momentum $(E_1 + E_2 + \ldots + E_n)/c^2, p_1 + p_2 + \ldots + p_n)$.

The relativistic law of energy-momentum conservation thus combines and generalizes in one relativistically invariant expression the separate conservation laws of prerelativistic physics: the conservation of mass, the conservation of momentum, and the conservation of energy. In fact, the law of conservation of mass becomes incorporated in the law of conservation of energy and is modified if the amount of energy exchanged is comparable with the rest mass energy of any of the particles.

For example, if a particle of mass M at rest decays into two particles the sum of whose rest masses $m_1 + m_2$ is smaller than M, then the two momenta p_1 and p_2 must be equal in magnitude and opposite in direction. The quantity $T = E - mc^2$ is the kinetic energy of the particle. In such a decay the initial kinetic energy is zero. Since the conservation of energy implies that in the process $Mc^2 = T_1 + T_2 + m_1c^2 + m_2c^2$, one speaks of the conversion of an amount $(M - m_1 - m_2)c^2$ of rest mass energy to kinetic energy. It is precisely this process that provides the large amount of energy available during nuclear fission, for example, in the spontaneous fission of the uranium-235 isotope. The opposite process occurs in nuclear fusion when two particles fuse to form a particle of smaller total rest mass. The difference $(m_1 + m_2 - M)$ multiplied by c^2 is called the binding energy. If the two initial particles are both at rest, a fourth particle is required to satisfy the conservation of energy and momentum. The rest mass of this fourth particle will not change, but it will acquire kinetic energy equal to the binding energy minus the kinetic energy of the fused particles. Perhaps the most important examples are the conversion of hydrogen to helium in the centre of stars, such as the Sun, and during thermonuclear reactions used in atomic bombs.

This article has so far dealt only with particles with non-vanishing rest mass whose velocities must always be less than that of light. One may always find an inertial reference frame with respect to which they are at rest and their energy in that frame equals mc^2. However, special relativity allows a generalization of classical ideas to include particles with vanishing rest masses that can move only with the velocity of light. Particles in nature that correspond to this possibility and that could not, therefore, be incorporated into the classical scheme are the photon, which is associated with the transmission of electromagnetic radiation, and—more speculatively—the graviton, which plays the same role with respect to gravitational waves as does the photon with respect to electromagnetic waves.

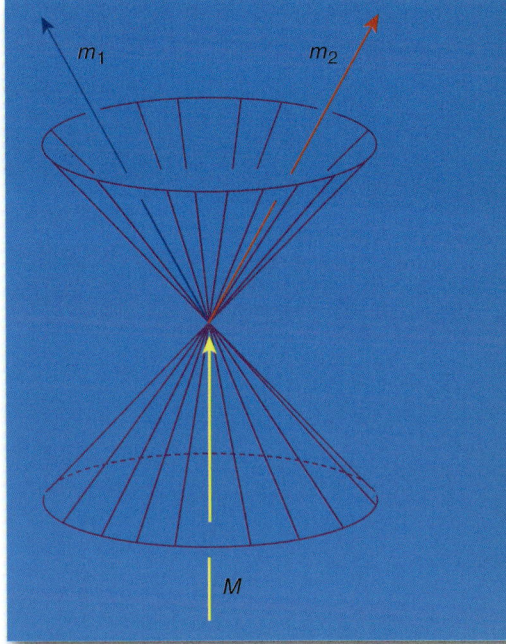

The decay of a particle of mass M into two particles the sum of whose rest masses is less than M. Copyright Encyclopædia Britannica; rendering for this edition by Rosen Educational Services

The velocity v of any particle in relativistic mechanics is given by $v = pc^2/E$, and the relation between energy E and momentum is $E^2 = m^2c^4 + p^2c^2$. Thus for massless particles $E = |p|c$ and the 4-momentum is given by $(|p|/c, p)$. It follows from the relativistic laws of energy and momentum conservation that, if a massless particle were to decay, it could do so only if the particles produced were all strictly massless and their momenta $p_1, p_2, \ldots p_n$ were all strictly aligned with the momentum p of the original massless particle. Since this is a situation of vanishing likelihood, it follows that strictly massless particles are absolutely stable.

It also follows that one or more massive particles cannot decay into a single massless particle, conserving both

The world lines of an electron (moving forward in time) and a positron (moving backward in time) that annihilate into two photons (see text). Copyright Encyclopædia Britannica; rendering for this edition by Rosen Educational Services

energy and momentum. They can, however, decay into two or more massless particles, and indeed this is observed in the decay of the neutral pion into photons and in the annihilation of an electron and a positron pair into photons. In the latter case, the world lines of the annihilating particles meet at the space-time event where they annihilate. Using the interpretation of Feynman and Stückelberg, one may view these two world lines as a single continuous world line with two portions, one moving forward in time and one moving backward in time. This interpretation plays an important role in the quantum theory of such processes.

Experimental Evidence for Special Relativity

Because relativistic changes are small at typical speeds for macroscopic objects, the confirmation of special relativity has relied on either the examination of subatomic bodies at high speeds or the measurement of small changes by sensitive instrumentation. For example, ultra-accurate clocks were placed on a variety of commercial airliners flying at one-millionth the speed of light. After two days of continuous flight, the time shown by the airborne clocks differed by fractions of a microsecond from that shown by a synchronized clock left on Earth, as predicted.

Larger effects are seen with elementary particles moving at speeds close to that of light. One such experiment involved muons, elementary particles created by cosmic rays in the Earth's atmosphere at an altitude of about 9 km (30,000 feet). At 99.8 percent of the speed of light, the muons should reach sea level in 31 microseconds, but measurements showed that it took only 2 microseconds. The reason is that, relative to the moving muons, the distance of 9 km contracted to 0.58 km (1,900 feet). Similarly, a relativistic mass increase has been confirmed in measurements on fast-moving elementary particles, where the change is large.

Such results leave no doubt that special relativity correctly describes the universe, although the theory is difficult to accept at a visceral level. Some insight comes from Einstein's comment that in relativity the limiting speed of light plays the role of an infinite speed. At infinite speed, light would traverse any distance in zero time. Similarly, according to the relativistic equations, an observer riding a light wave would see lengths contract to zero and clocks stop ticking as the universe approached him at the speed of light. Effectively, relativity replaces an infinite speed limit with the finite value of 3×10^8 metres per second.

CHAPTER 2
GENERAL RELATIVITY

Because Isaac Newton's law of gravity served so well in explaining the behaviour of the solar system, the question arises why it was necessary to develop a new theory of gravity. The answer is that Newton's theory violates special relativity, for it requires an unspecified "action at a distance" through which any two objects—such as the Sun and the Earth—instantaneously pull each other, no matter how far apart. However, instantaneous response would require the gravitational interaction to propagate at infinite speed, which is precluded by special relativity.

In practice, this is no great problem for describing our solar system, for Newton's law gives valid answers for objects moving slowly compared with light. Nevertheless, since Newton's theory cannot be conceptually reconciled with special relativity, Einstein turned to the development of general relativity as a new way to understand gravitation.

PRINCIPLE OF EQUIVALENCE

In order to begin building his theory, Einstein seized on an insight that came to him in 1907. As he explained in a lecture in 1922:

> *I was sitting on a chair in my patent office in Bern. Suddenly a thought struck me: If a man falls freely, he would not feel his weight. I was taken aback. This simple thought experiment made a deep impression on me. This led me to the theory of gravity.*

Einstein was alluding to a curious fact known in Newton's time: no matter what the mass of an object, it falls toward the Earth with the same acceleration (ignoring air resistance) of 9.8 metres per second squared. Newton explained this by postulating two types of mass: inertial mass, which resists motion and enters into his general laws of motion, and gravitational mass, which enters into his equation for the force of gravity. He showed that, if the two masses were equal, then all objects would fall with that same gravitational acceleration.

Einstein, however, realized something more profound. A person standing in an elevator with a broken cable feels weightless as the enclosure falls freely toward the Earth. The reason is that both he and the elevator accelerate downward at the same rate and so fall at exactly the same speed; hence, short of looking outside the elevator at his surroundings, he cannot determine that he is being pulled downward. In fact, there is no experiment he can do within a sealed falling elevator to determine that he is within a gravitational field. If he releases a ball from his hand, it will fall at the same rate, simply remaining where he releases it. And if he were to see the ball sink toward the floor, he could not tell if that was because he was at rest within a gravitational field that pulled the ball down or because a cable was yanking the elevator up so that its floor rose toward the ball.

Einstein expressed these ideas in his deceptively simple principle of equivalence, which is the basis of general relativity: on a local scale—meaning within a given system, without looking at other systems—it is impossible to distinguish between physical effects due to gravity and those due to acceleration.

In that case, continued Einstein's *Gedankenexperiment*, light must be affected by gravity. Imagine that the elevator has a hole bored straight through two opposite walls. When the elevator is at rest, a beam of light entering one hole travels in a straight line parallel to the floor and exits through the other hole. But if the elevator is accelerated upward, by the time the ray reaches the second hole, the opening has moved and is no longer aligned with the ray. As the passenger sees the light miss the second hole, he concludes that the ray has followed a curved path (in fact, a parabola).

If a light ray is bent in an accelerated system, then, according to the principle of equivalence, light should also be bent by gravity, contradicting the everyday expectation that light will travel in a straight line (unless it passes from one medium to another). If its path is curved by gravity, that must mean that "straight line" has a different meaning near a massive gravitational body such as a star than it does in empty space. This was a hint that gravity should be treated as a geometric phenomenon.

CURVED SPACE-TIME AND GEOMETRIC GRAVITATION

The singular feature of Einstein's view of gravity is its geometric nature. Whereas Newton thought that gravity was a force, Einstein showed that gravity arises from the shape of space-time. While this is difficult to visualize, there is an analogy that provides some insight—although it is only a guide, not a definitive statement of the theory.

The analogy begins by considering space-time as a rubber sheet that can be deformed. In any region distant from massive cosmic objects such as stars, space-time is uncurved—that is, the rubber sheet is absolutely flat. If one were to probe space-time in that region by sending out a ray of light or a test body, both the ray and the body would travel in perfectly straight lines, like a child's marble rolling across the rubber sheet.

General Relativity

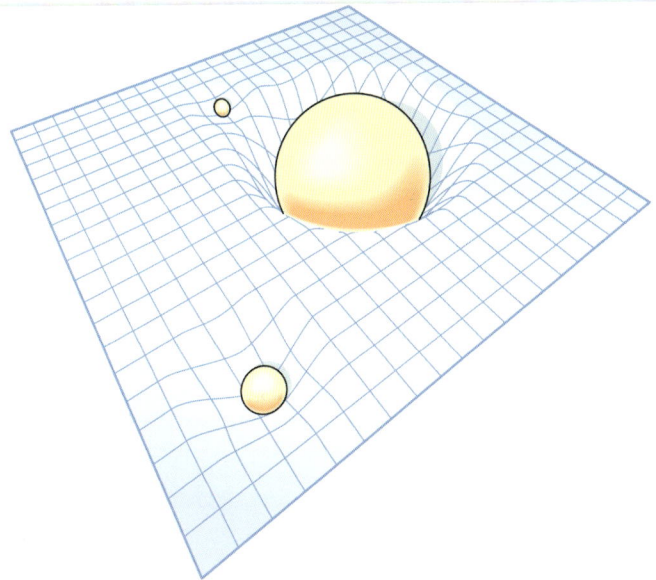

The four-dimensional space-time continuum itself is distorted in the vicinity of any mass, with the amount of distortion depending on the mass and the distance from the mass. Thus, relativity accounts for Newton's inverse square law of gravity through geometry and thereby does away with the need for any mysterious "action at a distance." Encyclopædia Britannica, Inc.

However, the presence of a massive body curves space-time, as if a bowling ball were placed on the rubber sheet to create a cuplike depression. In the analogy, a marble placed near the depression rolls down the slope toward the bowling ball as if pulled by a force. In addition, if the marble is given a sideways push, it will describe an orbit around the bowling ball, as if a steady pull toward the ball is swinging the marble into a closed path.

In this way, the curvature of space-time near a star defines the shortest natural paths, or geodesics — much as the shortest path between any two points on the Earth is not a straight line, which cannot be constructed on that curved surface, but the arc of a great circle route. In Einstein's theory, space-time geodesics define the deflection of light and the orbits of planets. As the American theoretical physicist John Wheeler put it, matter tells space-time how to curve, and space-time tells matter how to move.

THE MATHEMATICS OF GENERAL RELATIVITY

The rubber sheet analogy helps with visualization of space-time, but Einstein himself developed a complete quantitative theory that describes space-time through highly abstract mathematics. General relativity is expressed in a set of interlinked differential equations that define how the shape of space-time depends on the amount of matter (or, equivalently, energy) in the region. The solution of these so-called field equations can yield answers to different physical situations, including the behaviour of individual bodies and of the entire universe.

Cosmological Solutions

Einstein immediately understood that the field equations could describe the entire cosmos. In 1917 he modified the original version of his equations by adding what he called the "cosmological term." This represented a force that acted to make the universe expand, thus counteracting gravity, which tends to make the universe contract. The result was a static universe, in accordance with the best knowledge of the time.

In 1922, however, the Soviet mathematician Aleksandr Aleksandrovich Friedmann showed that the field equations predict a dynamic universe, which can either expand forever or go through cycles of alternating expansion and contraction. Einstein came to agree with this result and abandoned his cosmological term. Later work, notably pioneering measurements by the American astronomer

Edwin Hubble and the development of the big-bang model, has confirmed and amplified the concept of an expanding universe.

Black Holes

In 1916 the German astronomer Karl Schwarzschild used the field equations to calculate the gravitational effect of a single spherical body such as a star. If the mass is neither very large nor highly concentrated, the resulting calculation will be the same as that given by Newton's theory of gravity. Thus, Newton's theory is not incorrect; rather, it constitutes a valid approximation to general relativity under certain conditions.

Schwarzschild also described a new effect. If the mass is concentrated in a vanishingly small volume—a singularity—gravity will become so strong that nothing pulled into the surrounding region can ever leave. Even light cannot escape. In the rubber sheet analogy, it as if a tiny massive object creates a depression so steep that nothing can escape it. In recognition that this severe space-time distortion would be invisible—because it would absorb light and never emit any—it was dubbed a black hole.

In quantitative terms, Schwarzschild's result defines a sphere that is centred at the singularity and whose radius depends on the density of the enclosed mass. Events within the sphere are forever isolated from the remainder of the universe; for this reason, the Schwarzschild radius is called the event horizon.

EXPERIMENTAL EVIDENCE FOR GENERAL RELATIVITY

Soon after the theory of general relativity was published in 1916, the English astronomer Arthur Eddington

In 1919 observation of a solar eclipse confirmed Einstein's prediction that light is bent in the presence of mass. This experimental support for his general theory of relativity garnered him instant worldwide acclaim. Encyclopædia Britannica, Inc.

considered Einstein's prediction that light rays are bent near a massive body, and he realized that it could be verified by carefully comparing star positions in images of the Sun taken during a solar eclipse with images of the same region of space taken when the Sun was in a different portion of the sky. Verification was delayed by World War I, but in 1919 an excellent opportunity presented itself with an especially long total solar eclipse, in the vicinity of the bright Hyades star cluster, that was visible from northern Brazil to the African coast. Eddington led one expedition to Príncipe, an island off the African coast, and Andrew Crommelin of the Royal Greenwich Observatory led a second expedition to Sobral, Brazil. After carefully comparing photographs from both expeditions with reference photographs of the Hyades, Eddington declared that the starlight had been deflected about 1.75 seconds of arc, as predicted by general relativity. (The same effect produces gravitational lensing, where a massive cosmic object focuses light from another object beyond it to produce a distorted or magnified image. The astronomical discovery of gravitational lenses in 1979 gave additional support for general relativity.)

Further evidence came from the planet Mercury. In the 19th century, it was found that Mercury does not return to exactly the same spot every time it completes its elliptical orbit. Instead, the ellipse rotates slowly in space, so that on each orbit the perihelion—the point of closest approach to the Sun—moves to a slightly different angle. Newton's law of gravity could not explain this perihelion shift, but general relativity gave the correct orbit.

Another confirmed prediction of general relativity is that time dilates in a gravitational field, meaning that clocks run slower as they approach the mass that is producing the field. This has been measured directly and also through the gravitational redshift of light. Time dilation causes light to vibrate at a lower frequency within a gravitational field; thus, the light is shifted toward a longer wavelength—that is, toward the red. Other measurements have verified the equivalence principle by showing that inertial and gravitational mass are precisely the same.

UNCONFIRMED PREDICTIONS OF GENERAL RELATIVITY

Although experiment and observation support general relativity, not all of its predictions have been realized. These include such phenomena as gravity waves and wormholes.

Gravitational Waves

The most striking unconfirmed prediction is that of gravitational waves, which replace Newton's instantaneous "action at a distance"; that is, general relativity predicts that the "wrinkles" in space-time curvature that represent gravity propagate at the speed of light.

Superficially, there are many similarities between gravity and electromagnetism. For example, Newton's law for the gravitational force between two point masses and Coulomb's law for the electric force between two point charges indicate that both forces vary as the inverse square of the separation distance. Yet in Scottish physicist James Clerk Maxwell's theory for electromagnetism, accelerated charges emit signals (electromagnetic radiation) that travel at the speed of light, whereas in Newton's theory of gravitation accelerated masses transmit information (action at a distance) that travels at infinite speed. This dichotomy is repaired by Einstein's theory of gravitation, wherein accelerated masses also produce signals (gravitational waves) that travel only at the speed of light. And, just as electromagnetic waves can make their presence known by the pushing to and fro of electrically charged bodies, so too should gravitational waves be detected, in principle, by the tugging to and fro of massive bodies. However, because the coupling of gravitational forces to masses is intrinsically much weaker than the coupling of electromagnetic forces to charges, the generation and detection of gravitational radiation are much more difficult than those of electromagnetic radiation. Indeed, since the time of Einstein's discovery of general relativity in 1916, there has yet to be a single instance of the detection of gravitational waves that is direct and undisputed.

Nevertheless, there are strong grounds for believing that such radiation exists. The most convincing concerns radio-timing observations of a pulsar, PSR 1913+16, located in a binary star system with an orbital period of 7.75 hours. This object, discovered in 1974, has a pulse period of about 59 milliseconds that varies by about one part in 1,000 every 7.75 hours. Interpreted as Doppler shifts, these variations imply orbital velocities on the order of 1/1,000 the speed of light. The nonsinusoidal shape of the velocity curve with time allows a deduction that the orbit is quite noncircular (indeed, it is an ellipse of eccentricity 0.62 whose long axis precesses in space by 4.2° per year). It is now believed that the system is composed of two neutron stars, each having a mass of about 1.4 solar masses, with a semimajor axis separation of only 2.8 solar radii. According to Einstein's theory of general relativity, such a system ought to be losing orbital energy through the radiation of gravitational waves at a rate that would cause them to spiral together on a timescale of about 3×10^8 years. The observed decrease in the orbital period in the years since the discovery of the binary pulsar does indeed indicate that the two stars are spiraling toward one another at exactly the predicted rate. Gravitational radiation is the only known means by which that could happen. (American physicists Russell Hulse and Joseph H. Taylor, Jr., won the Nobel Prize for Physics in 1993 for their discovery of PSR 1913+16.)

The implosion of the core of a massive star to form a neutron star prior to a supernova explosion, if it takes place in a nonspherically symmetric way, ought to provide a powerful burst of gravitational radiation. Simple estimates yield the release of a fraction of the mass-energy deficit, roughly 10^{53} ergs, with the radiation primarily coming out at wave periods between the vibrational period of the neutron star, approximately 0.3 millisecond, and the gravitational-radiation damping time, about 300 milliseconds.

Three types of detectors have been designed to look for gravitational radiation, which is expected to be very weak. The changes of curvature of space-time would correspond to a dilation in one direction and a contraction at right angles to that direction. One scheme, first tried out about 1960, employs a massive cylinder that might be set in mechanical oscillation by a gravitational signal. The authors of this apparatus argued that signals had been detected, but their claim has not been substantiated. In later developments the cylinder has been cooled by liquid helium, and great attention has been paid to possible disturbances. In a second scheme an optical interferometer is set up with freely suspended reflectors at the ends of long paths that are at right angles to each other. Shifts of interference fringes corresponding to an increase in length of one arm and a decrease in the other would indicate the passage of gravitational waves. One such interferometer is the Laser Interferometer Gravitational-Wave Observatory (LIGO), which consists of two interferometers with arm lengths of 4 km (2 miles), one in Hanford, Wash., and the other in Livingston, La. LIGO and other interferometers have not yet directly observed gravitational radiation. A third scheme, the Laser Interferometer Space Antenna (LISA), is planned that uses three separate, but not independent, interferometers installed in three spacecraft located at the corners of a triangle with sides of some 5 million km (3 million miles).

Black Holes and Wormholes

No human technology could compact matter sufficiently to make black holes, but they may occur as final steps in the life cycle of stars. After millions or billions of years, a star uses up all of its hydrogen and other elements that produce energy through nuclear fusion. With its nuclear furnace banked, the star no longer maintains an internal pressure to expand, and gravity is left unopposed to pull inward and compress the star. For stars above a certain mass, this gravitational collapse will in principle produce a black hole containing several times the mass of the Sun. In other cases, the gravitational collapse of huge dust clouds may create supermassive black holes containing millions or billions of solar masses.

Astrophysicists have found several cosmic objects that appear to contain a dense concentration of mass in a small volume. These strong candidates for black holes include one at the centre of the Milky Way Galaxy and certain binary stars that emit X-rays as they orbit each other. However, the definitive signature of a black hole, the event horizon, has not been observed.

The theory of black holes has led to another predicted entity, a wormhole. This is a solution of the field equations that resembles a tunnel between two black holes or other points in space-time. Such a tunnel would provide a shortcut between its end points. In analogy, consider an ant walking across a flat sheet of paper from point A to point B. If the paper is curved through the third dimension, so that A and B overlap, the ant can step directly from one point to the other, thus avoiding a long trek.

The possibility of short-circuiting the enormous distances between stars makes wormholes attractive for space travel. Because the tunnel links moments in time as well as locations in space, it also has been argued that a wormhole would allow travel into the past. However, wormholes are intrinsically unstable. While exotic stabilization schemes have been proposed, there is as yet no evidence that these can work or indeed that wormholes exist.

APPLICATIONS OF RELATIVISTIC IDEAS

Although relativistic effects are negligible in ordinary life, relativistic ideas appear in a range of areas from fundamental science to civilian and military technology.

Elementary Particles

The relationship $E = mc^2$ is essential in the study of subatomic particles. It determines the energy required to create particles or to convert one type into another and the energy released when a particle is annihilated. For example, two photons, each of energy E, can collide to form two particles, each with mass $m = E/c^2$. This pair-production process is one step in the early evolution of the universe, as described in the big-bang model.

Particle Accelerators

Knowledge of elementary particles comes primarily from particle accelerators. These machines raise subatomic particles, usually electrons or protons, to nearly the speed of light. When these energetic bullets smash into selected targets, they elucidate how subatomic particles interact and often produce new species of elementary particles.

Particle accelerators could not be properly designed without special relativity. In the type called an electron synchrotron, for instance, electrons gain energy as they traverse a huge circular raceway. At barely below the speed of light, their mass is thousands of times larger than their rest mass. As a result, the magnetic field used to hold the electrons in circular orbits must be thousands of times stronger than if the mass did not change.

Fission and Fusion: Bombs and Stellar Processes

Energy is released in two kinds of nuclear processes. In nuclear fission a heavy nucleus, such as uranium, splits into two lighter nuclei; in nuclear fusion two light nuclei combine into a heavier one. In each process the total final mass is less than the starting mass. The difference appears as energy according to the relation $E = \Delta mc^2$, where Δm is the mass deficit.

Fission is used in atomic bombs and in reactors that produce power for civilian and military applications. The fusion of hydrogen into helium is the energy source in stars and provides the power of a hydrogen bomb. Efforts are now under way to develop controllable hydrogen fusion as a clean, abundant power source.

The Global Positioning System

The global positioning system (GPS) depends on relativistic principles. A GPS receiver determines its location on Earth's surface by processing radio signals from four or more satellites. The distance to each satellite is calculated as the product of the speed of light and the time lag between transmission and reception of the signal. However, Earth's gravitational field and the motion of the satellites cause time-dilation effects, and Earth's rotation also has relativistic implications. Hence, GPS technology includes relativistic corrections that enable positions to be calculated to within several centimetres.

Cosmology

To derive his 1917 cosmological model, Einstein made three assumptions that lay outside the scope of his equations. The first was to suppose that the universe is homogeneous and isotropic in the large (i.e., the same everywhere on average at any instant in time), an assumption that the English astrophysicist Edward A. Milne later elevated to an entire philosophical outlook by naming it the cosmological principle. Given the success of the Copernican revolution, this outlook is a natural one. Newton himself had it implicitly in mind when he took the initial state of the universe to be everywhere the same before it developed "ye Sun and Fixt stars."

The second assumption was to suppose that this homogeneous and isotropic universe had a closed spatial geometry. As described above, the total volume of a three-dimensional space with uniform positive curvature would be finite but possess no edges or boundaries (to be consistent with the first assumption).

The third assumption made by Einstein was that the universe as a whole is static—i.e., its large-scale properties do not vary with time. This assumption, made before Hubble's observational discovery of the expansion of the universe, was also natural; it was the simplest approach, as Aristotle had discovered, if one wishes to avoid a discussion of a creation event. Indeed, the philosophical attraction of the notion that the universe on average is not only homogeneous and isotropic in space but also constant in time was so appealing that a school of English cosmologists—Hermann Bondi, Fred Hoyle, and Thomas Gold—would call it the perfect cosmological principle and carry its implications in the 1950s to the ultimate refinement in the so-called steady-state theory.

To his great chagrin Einstein found in 1917 that with his three adopted assumptions, his equations of general relativity—as originally written down—had no meaningful solutions. To obtain a solution, Einstein realized that he had to add to his equations an extra term, which came to be called the cosmological constant. If one speaks in Newtonian terms, the cosmological constant could be interpreted as a repulsive force of unknown origin that could exactly balance the attraction of gravitation of all the matter in Einstein's closed universe and keep it from moving. The inclusion of such a term in a more general context, however, meant that the universe in the absence of any mass-energy (i.e., consisting of a vacuum) would not have a space-time structure that was flat (i.e., would not have satisfied the dictates of special relativity exactly). Einstein was prepared to make such a sacrifice only very reluctantly, and, when he later learned of Hubble's discovery of the expansion of the universe and realized that he could have predicted it had he only had more faith in the original form of his equations, he regretted the introduction of the cosmological constant as the "biggest blunder" of his life. Ironically, observations of distant supernovas have shown the existence of dark energy, a repulsive force that is the dominant component of the universe.

It was also in 1917 that the Dutch astronomer Willem de Sitter recognized that he could obtain a static cosmological model differing from Einstein's simply by removing all matter. The solution remains stationary essentially because there is no matter to move about. If some test particles are reintroduced into the model, the cosmological term would propel them away from each other. Astronomers now began to wonder if this effect might not underlie the recession of the spiral galaxies.

In 1922 Aleksandr A. Friedmann, a Russian meteorologist and mathematician, and in 1927 Georges Lemaître, a Belgian cleric, independently discovered solutions to Einstein's equations that contained realistic amounts of matter. These evolutionary models correspond to big bang cosmologies. Friedmann and Lemaître adopted Einstein's assumption of spatial homogeneity and isotropy (the cosmological principle). They rejected, however, his assumption of time independence and considered both positively curved spaces ("closed" universes) as well as negatively curved spaces ("open" universes). The difference between the approaches of Friedmann and Lemaître

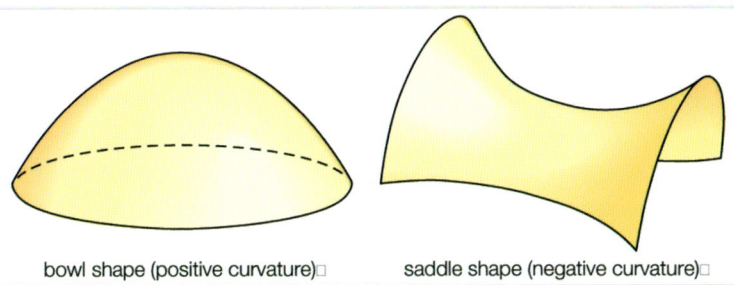

Intrinsic curvature of a surface. Encyclopædia Britannica, Inc.

is that the former set the cosmological constant equal to zero, whereas the latter retained the possibility that it might have a nonzero value. To simplify the discussion, only the Friedmann models are considered here.

The decision to abandon a static model meant that the Friedmann models evolve with time. As such, neighbouring pieces of matter have recessional (or contractional) phases when they separate from (or approach) one another with an apparent velocity that increases linearly with increasing distance. Friedmann's models thus anticipated Hubble's law before it had been formulated on an observational basis. It was Lemaître, however, who had the good fortune of deriving the results at the time when the recession of the galaxies was being recognized as a fundamental cosmological observation, and it was he who clarified the theoretical basis for the phenomenon.

The geometry of space in Friedmann's closed models is similar to that of Einstein's original model; however, there is a curvature to time as well as one to space. Unlike Einstein's model, where time runs eternally at each spatial point on an uninterrupted horizontal line that extends infinitely into the past and future, there is a beginning and end to time in Friedmann's version of a closed universe when material expands from or is recompressed to infinite densities. These instants are called the instants of the "big bang" and the "big squeeze," respectively. The global space-time diagram for the middle half of the expansion-compression phases can be depicted as a barrel lying on its side. The space axis corresponds again to any one direction in the universe, and it wraps around the barrel. Through each spatial point runs a time axis that extends along the length of the barrel on its (space-time) surface. Because the barrel is curved in both space and time, the little squares in the grid of the curved sheet of graph paper marking the space-time surface are of nonuniform size, stretching to become bigger when the barrel broadens (universe expands) and shrinking to become smaller when the barrel narrows (universe contracts).

It should be remembered that only the surface of the barrel has physical significance; the dimension off the surface toward the axle of the barrel represents the fourth spatial dimension, which is not part of the real three-dimensional world. The space axis circles the barrel and closes upon itself after traversing a circumference equal to $2\pi R$, where R, the radius of the universe (in the fourth dimension), is now a function of the time t. In a closed Friedmann model, R starts equal to zero at time $t = 0$ (not shown in barrel diagram), expands to a maximum value at time $t = t_m$ (the middle of the barrel), and recontracts to zero (not shown) at time $t = 2t_m$, with the value of t_m dependent on the total amount of mass that exists in the universe.

Imagine now that galaxies reside on equally spaced tick marks along the space axis. Each galaxy on average does not move spatially with respect to its tick mark in the spatial (ringed) direction but is carried forward horizontally by the march of time. The total number of galaxies on the spatial ring is conserved as time changes, and therefore their average spacing increases or decreases as the total circumference $2\pi R$ on the ring increases or decreases (during the expansion or contraction phases). Thus, without in a sense actually moving in the spatial direction, galaxies can be carried apart by the expansion of space itself. From this point of view, the recession of galaxies is not a "velocity" in the usual sense of the word. For example, in a closed Friedmann model, there could be galaxies that started, when R was small, very close to the Milky Way system on the opposite side of the universe. Now, 10^{10} years later, they are still on the opposite side of the universe but at a distance much greater than 10^{10} light-years away. They reached those distances without ever having had to move (relative to any local observer) at speeds faster than light—indeed, in a sense without having had to move at all. The separation rate of nearby galaxies can be thought of as a velocity without confusion in the sense of Hubble's law, if one wants, but only if the inferred velocity is much less than the speed of light.

On the other hand, if the recession of the galaxies is not viewed in terms of a velocity, then the cosmological redshift cannot be viewed as a Doppler shift. How, then, does it arise? The answer is contained in the barrel diagram when one notices that, as the universe expands, each small cell in the space-time grid also expands. Consider the propagation of electromagnetic radiation whose wavelength initially spans exactly one cell length (for simplicity of discussion), so that its head lies at a vertex and its tail at one vertex back. Suppose an elliptical galaxy emits such a wave at some time t_1. The head of the wave propagates from corner to corner on the little square grids that look locally flat, and the tail propagates from corner to corner one vertex back. At a later time t_2, a spiral galaxy begins to intercept the head of the wave. At time t_2, the tail is still one vertex back, and therefore the wave train, still containing one wavelength, now spans one current spatial grid spacing. In other words, the wavelength has grown in direct proportion to the linear expansion factor of the universe. Since the same conclusion would have held if n wavelengths had been involved instead of one, all electromagnetic radiation from a given object will show the same cosmological redshift if the universe (or, equivalently, the average spacing between galaxies) was smaller at the epoch of transmission than at the epoch of reception. Each wavelength will have been stretched in direct proportion to the expansion of the universe in between.

A nonzero peculiar velocity for an emitting galaxy with respect to its local cosmological frame can be taken into account by Doppler-shifting the emitted photons before applying the cosmological redshift factor; i.e., the observed redshift would be a product of two factors. When the observed redshift is large, one usually assumes that the dominant contribution is of cosmological origin. When this assumption is valid, the redshift is a monotonic function of both distance and time during the expansional phase of any cosmological model. Thus, astronomers often use the redshift z as a shorthand indicator of both distance and elapsed time. Following from this, the statement "object X lies at $z = a$" means that "object X lies at a distance associated with redshift a"; the statement "event Y occurred at redshift $z = b$" means that "event Y occurred a time ago associated with redshift b."

The open Friedmann models differ from the closed models in both spatial and temporal behaviour. In an open universe the total volume of space and the number of galaxies contained in it are infinite. The three-dimensional spatial geometry is one of uniform negative curvature in the sense that, if circles are drawn with very large lengths of string, the ratio of circumferences to lengths of string are greater than 2π. The temporal history begins again with expansion from a big bang of infinite density, but now the expansion continues indefinitely, and the average density of matter and radiation in the universe would eventually become vanishingly small. Time in such a model has a beginning but no end.

In 1932 Einstein and de Sitter proposed that the cosmological constant should be set equal to zero, and they derived a homogeneous and isotropic model that provides the separating case between the closed and open Friedmann models; i.e., Einstein and de Sitter assumed that the spatial curvature of the universe is neither positive nor negative but rather zero. The spatial geometry of the Einstein–de Sitter universe is Euclidean (infinite total

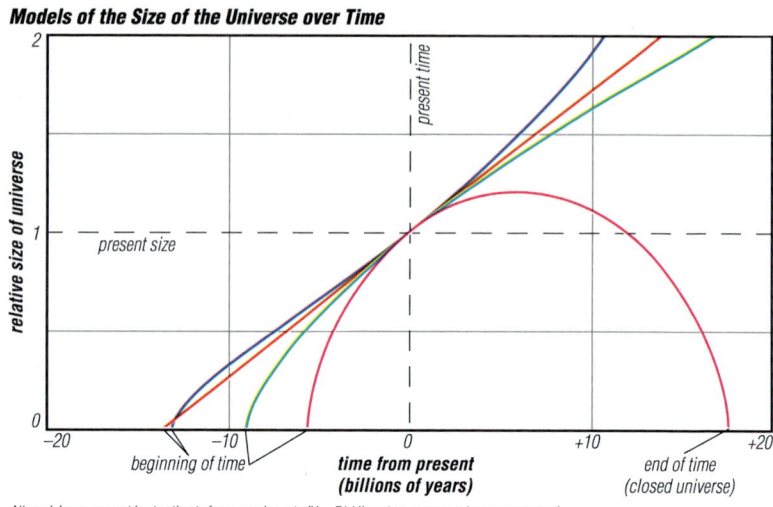

How the relative size of the universe changes with time in four different models. The red line shows a universe devoid of matter, with constant expansion. Pink shows a collapsing universe, with six times the critical density of matter. Green shows a model favoured until 1998, with exactly the critical density and a universe 100 percent matter. Blue shows the currently favoured scenario, with exactly the critical density, of which 27 percent is visible and dark matter and 73 percent is dark energy. Encyclopædia Britannica, Inc.

volume), but space-time is not globally flat (i.e., not exactly the space-time of special relativity). Time again commences with a big bang and the galaxies recede forever, but the recession rate (Hubble's "constant") asymptotically coasts to zero as time advances to infinity. Because the geometry of space and the gross evolutionary properties are uniquely defined in the Einstein–de Sitter model, many people with a philosophical bent long considered it the most fitting candidate to describe the actual universe.

The different separation behaviours of galaxies at large timescales in the Friedmann closed and open models and the Einstein–de Sitter model allow a different classification scheme than one based on the global structure of space-time. The alternative way of looking at things is in terms of gravitationally bound and unbound systems: closed models where galaxies initially separate but later come back together again represent bound universes; open models where galaxies continue to separate forever represent unbound universes; the Einstein–de Sitter model where galaxies separate forever but slow to a halt at infinite time represents the critical case.

The advantage of this alternative view is that it focuses attention on local quantities where it is possible to think in the simpler terms of Newtonian physics—attractive forces, for example. In this picture it is intuitively clear that the feature that should distinguish whether or not gravity is capable of bringing a given expansion rate to a halt depends on the amount of mass (per unit volume) present. This is indeed the case; the Newtonian and relativistic formalisms give the same criterion for the critical, or closure, density (in mass equivalent of matter and radiation) that separates closed or bound universes from open or unbound ones. If Hubble's constant at the present epoch is denoted as H_o, then the closure density (corresponding to an Einstein–de Sitter model) equals $3H_o^2/8\pi G$, where G is the universal gravitational constant in both Newton's and Einstein's theories of gravity. The numerical value of Hubble's constant H_o is 22 kilometres per second per million light-years; the closure density then equals 10^{-29} gram per cubic centimetre, the equivalent of about six hydrogen atoms on average per cubic metre of cosmic space. If the actual cosmic average is greater than this value, the universe is bound (closed) and, though currently expanding, will end in a crush of unimaginable proportion. If it is less, the universe is unbound (open) and will expand forever. The result is intuitively plausible since the smaller the mass density, the smaller the role for gravitation, so the more the universe will approach free expansion (assuming that the cosmological constant is zero).

The mass in galaxies observed directly, when averaged over cosmological distances, is estimated to be only a few percent of the amount required to close the universe. The amount contained in the radiation field (most of which is in the cosmic microwave background) contributes negligibly to the total at present. If this were all, the universe would be open and unbound. However, the dark matter that has been deduced from various dynamic arguments is about 23 percent of the universe, and dark energy supplies the remaining amount, bringing the total average mass density up to 100 percent of the closure density.

RELATIVITY, QUANTUM THEORY, AND UNIFIED THEORIES

Cosmic behaviour on the biggest scale is described by general relativity. Behaviour on the subatomic scale is described by quantum mechanics, which began with the work of the German physicist Max Planck in 1900 and treats energy and other physical quantities in discrete units called quanta. A central goal of physics has been to combine relativity theory and quantum theory into an overarching "theory of everything" describing all physical phenomena. Quantum theory explains electromagnetism and the strong and weak forces, but a quantum description of the remaining fundamental force of gravity has not been achieved.

After Einstein developed relativity, he unsuccessfully sought a so-called unified field theory with a space-time geometry that would encompass all the fundamental forces. Other theorists have attempted to merge general relativity with quantum theory, but the two approaches treat forces in fundamentally different ways. In quantum theory, forces arise from the interchange of certain elementary particles, not from the shape of space-time.

Furthermore, quantum effects are thought to cause a serious distortion of space-time at an extremely small scale called the Planck length, which is much smaller than the size of elementary particles. This suggests that quantum gravity cannot be understood without treating space-time at unheard-of scales.

Although the connection between general relativity and quantum mechanics remains elusive, some progress has been made toward a fully unified theory. In the 1960s, the electroweak theory provided partial unification, showing a common basis for electromagnetism and the weak force within quantum theory. Recent research suggests that superstring theory, in which elementary particles are represented not as mathematical points but as extremely small strings vibrating in 10 or more dimensions, shows promise for supporting complete unification, including gravitation. However, until confirmed by experimental results, superstring theory will remain an untested hypothesis.

INTELLECTUAL AND CULTURAL IMPACT OF RELATIVITY

The impact of relativity has not been limited to science. Special relativity arrived on the scene at the beginning of the 20th century, and general relativity became widely known after World War I—eras when a new sensibility of "modernism" was becoming defined in art and literature. In addition, the confirmation of general relativity provided by the solar eclipse of 1919 received wide publicity. Einstein's 1921 Nobel Prize for Physics (awarded for his work on the photon nature of light), as well as the popular perception that relativity was so complex that few could grasp it, quickly turned Einstein and his theories into cultural icons.

The ideas of relativity were widely applied—and misapplied—soon after their advent. Some thinkers interpreted the theory as meaning simply that all things are relative, and they employed this concept in arenas distant from physics. The Spanish humanist philosopher and essayist José Ortega y Gasset, for instance, wrote in *The Modern Theme* (1923),

> *The theory of Einstein is a marvelous proof of the harmonious multiplicity of all possible points of view. If the idea is extended to morals and aesthetics, we shall come to experience history and life in a new way.*

The revolutionary aspect of Einstein's thought was also seized upon, as by the American art critic Thomas Craven, who in 1921 compared the break between classical and modern art to the break between Newtonian and Einsteinian ideas about space and time.

Some saw specific relations between relativity and art arising from the idea of a four-dimensional space-time continuum. In the 19th century, developments in geometry led to popular interest in a fourth spatial dimension, imagined as somehow lying at right angles to all three of the ordinary dimensions of length, width, and height. Edwin Abbott's *Flatland* (1884) was the first popular presentation of these ideas. Other works of fantasy that followed spoke of the fourth dimension as an arena apart from ordinary existence.

Einstein's four-dimensional universe, with three spatial dimensions and one of time, is conceptually different from four spatial dimensions. But the two kinds of four-dimensional world became conflated in interpreting the new art of the 20th century. Early Cubist works by Pablo Picasso that simultaneously portrayed all sides of their subjects became connected with the idea of higher dimensions in space, which some writers attempted to relate to relativity. In 1949, for example, the art historian

Paul LaPorte wrote that "the new pictorial idiom created by [C]ubism is most satisfactorily explained by applying to it the concept of the space-time continuum." Einstein specifically rejected this view, saying, "This new artistic 'language' has nothing in common with the Theory of Relativity." Nevertheless, some artists explicitly explored Einstein's ideas. In the new Soviet Union of the 1920s, for example, the poet and illustrator Vladimir Mayakovsky, a founder of the artistic movement called Russian Futurism, or Suprematism, hired an expert to explain relativity to him.

The widespread general interest in relativity was reflected in the number of books written to elucidate the subject for nonexperts. Einstein's popular exposition of special and general relativity appeared almost immediately, in 1916. Other scientists, such as the Russian mathematician Aleksandr Friedmann and the British astronomer Arthur Eddington, wrote popular books on the subjects in the 1920s. Such books continued to appear decades later.

When relativity was first announced, the public was typically awestruck by its complexity, a justified response to the intricate mathematics of general relativity. But the abstract, nonvisceral nature of the theory also generated reactions against its apparent violation of common sense. These reactions included a political undertone; in some quarters, it was considered undemocratic to present or support a theory that could not be immediately understood by the common person.

In contemporary usage, general culture has accepted the ideas of relativity—the impossibility of faster-than-light travel, $E = mc^2$, time dilation and the twin paradox, the expanding universe, and black holes and wormholes—to the point where they are immediately recognized in the media and provide plot devices for works of science fiction. Some of these ideas have gained meaning beyond their strictly scientific ones; in the business world, for instance, "black hole" can mean an unrecoverable financial drain.

In 1925 the British philosopher Bertrand Russell, in his *ABC of Relativity*, suggested that Einstein's work would lead to new philosophical concepts. Relativity has indeed had a great effect on philosophy, illuminating some issues that go back to the ancient Greeks. The idea of the ether, invoked in the late 19th century to carry light waves, harks back to Aristotle. He divided the world into earth, air, fire, and water, with the ether (aether) as the fifth element representing the pure celestial sphere. The Michelson-Morley experiment and relativity eliminated the last vestiges of this idea.

Relativity also changed the meaning of geometry as it was developed in Euclid's *Elements* (c. 300 BCE). Euclid's system relied on the axiom "a straight line is the shortest distance between two points," among others that seemed self-evidently true. Straight lines also played a special role in Euclid's *Optics* as the paths followed by light rays. To philosophers such as the German Immanuel Kant, Euclid's straight-line axiom represented a deep level of truth. But general relativity makes it possible scientifically to examine space like any other physical quantity—that is, to investigate Euclid's premises. It is now known that space-time is curved near stars; no straight lines exist there, and light follows curved geodesics. Like Newton's law of gravity, Euclid's geometry correctly describes reality under certain conditions, but its axioms are not absolutely fundamental and universal, for the cosmos includes non-Euclidean geometries as well.

Considering its scientific breadth, its recasting of people's view of reality, its ability to describe the entire universe, and its influence outside science, Einstein's relativity stands among the most significant and influential of scientific theories.

CHAPTER 3
QUANTUM MECHANICS:
CONCEPTS

Quantum mechanics is the science dealing with the behaviour of matter and light on the atomic and subatomic scale. It attempts to describe and account for the properties of molecules and atoms and their constituents—electrons, protons, neutrons, and other more esoteric particles such as quarks and gluons. These properties include the interactions of the particles with one another and with electromagnetic radiation (i.e., light, X-rays, and gamma rays).

HISTORICAL BASIS OF QUANTUM THEORY

At a fundamental level, both radiation and matter have characteristics of particles and waves. The gradual recognition by scientists that radiation has particle-like properties and that matter has wavelike properties provided the impetus for the development of quantum mechanics. Influenced by Newton, most physicists of the 18th century believed that light consisted of particles, which they called corpuscles. From about 1800, evidence began to accumulate for a wave theory of light. At about this time Thomas Young showed that, if monochromatic light passes through a pair of slits, the two emerging beams interfere, so that a fringe pattern of alternately bright and dark bands appears on a screen. The bands are readily explained by a wave theory of light. According to the theory, a bright band is produced when the crests (and troughs) of the waves from the two slits arrive together at the screen; a dark band is produced when the crest of one wave arrives at the same time as the trough of the other, and the effects of the two light beams cancel. Beginning in 1815, a series of experiments by Augustin-Jean Fresnel of France and others showed that, when a parallel beam of light passes through a single slit, the emerging beam is no longer parallel but starts to diverge; this phenomenon is known as diffraction. Given the wavelength of the light and the geometry of the apparatus (i.e., the separation and widths of the slits and the distance from the slits to the screen), one can use the wave theory to calculate the expected pattern in each case; the theory agrees precisely with the experimental data.

EARLY DEVELOPMENTS

By the end of the 19th century, physicists almost universally accepted the wave theory of light. However, though the ideas of classical physics explain interference and diffraction phenomena relating to the propagation of light, they do not account for the absorption and emission of light.

Planck's Radiation Law

All bodies radiate electromagnetic energy as heat; in fact, a body emits radiation at all wavelengths. The energy radiated at different wavelengths is a maximum at a wavelength that depends on the temperature of the body; the hotter the body, the shorter the wavelength for maximum radiation. Attempts to calculate the energy distribution for the radiation from a blackbody using classical ideas were unsuccessful. (A blackbody is a hypothetical ideal body or surface that absorbs and reemits all radiant energy falling on it.) One formula, proposed by Wilhelm Wien of Germany, did not agree with observations at long wavelengths, and another, proposed by Lord Rayleigh (John William Strutt) of England, disagreed with those at short wavelengths.

In 1900 the German theoretical physicist Max Planck made a bold suggestion. He assumed that the radiation energy is emitted, not continuously, but rather in discrete packets called quanta. For light of a given wavelength, the magnitude of all the quanta emitted or absorbed is the same in both energy and momentum. These particle-like packets of light are called photons, a term also applicable to quanta of other forms of electromagnetic energy such as X-rays and gamma rays.

The energy E of the quantum is related to the frequency v by $E = hv$. The quantity h, now known as the Planck constant, is a universal constant with the approximate value in metre-kilogram-second units of $6.6260669 \times 10^{-34}$ joule·second, with a standard uncertainty of $0.00000033 \times 10^{-34}$ joule·second. The dimension of Planck's constant is the product of energy multiplied by time, a quantity called action. Planck's constant is often defined, therefore, as the elementary quantum of action. Planck showed that the calculated energy spectrum then agreed with observation over the entire wavelength range.

Einstein and the Photoelectric Effect

In 1905 Einstein extended Planck's hypothesis to explain the photoelectric effect, which is the emission of electrons by a metal surface when it is irradiated by light or more-energetic photons. The kinetic energy of the emitted electrons depends on the frequency v of the radiation, not on its intensity; for a given metal, there is a threshold frequency v_o below which no electrons are emitted. Furthermore, emission takes place as soon as the light shines on the surface; there is no detectable delay. Einstein showed that these results can be explained by two assumptions: (1) that light is composed of corpuscles or photons, the energy of which is given by Planck's relationship, and (2) that an atom in the metal can absorb either a whole photon or nothing. Part of the energy of the absorbed photon frees an electron, which requires a fixed energy W, known as the work function of the metal; the rest is converted into the kinetic energy $m_e u^2/2$ of the emitted electron (m_e is the mass of the electron and u is its velocity). Thus, the energy relation is

$$hv = W + \frac{m_e u^2}{2}$$

If v is less than v_o, where $hv_o = W$, no electrons are emitted. Not all the experimental results mentioned above were known in 1905, but all Einstein's predictions have been verified since.

Bohr's Theory of the Atom

A major contribution to the subject was made by Niels Bohr of Denmark, who applied the quantum hypothesis to atomic spectra in 1913. The spectra of light emitted by gaseous atoms had been studied extensively since the mid-19th century. It was found that radiation from gaseous atoms at low pressure consists of a set of discrete wavelengths. This is quite unlike the radiation from a solid, which is distributed over a continuous range of wavelengths. The set of discrete wavelengths from gaseous atoms is known as a line spectrum, because the radiation (light) emitted consists of a series of sharp lines. The wavelengths of the lines are characteristic of the element and may form extremely complex patterns. The simplest spectra are those of atomic hydrogen and the alkali atoms (e.g., lithium, sodium, and potassium). For hydrogen, the wavelengths λ are given by the empirical formula

$$\frac{1}{\lambda} = R_\infty \left[\frac{1}{m^2} - \frac{1}{n^2} \right],$$

where m and n are positive integers with $n > m$ and R_∞, known as the Rydberg constant, has the value 1.097373157 × 10⁷ per metre, or 109,737.3157 per centimetre. When used in this form in the mathematical description of series of spectral lines, the result is the number of waves per unit length, or the wave numbers. Multiplication by the speed of light yields the frequencies of the spectral lines. For a given value of m, the lines for varying n form a series. The lines for $m = 1$, the Lyman series, lie in the ultraviolet part of the spectrum; those for $m = 2$, the Balmer series, lie in the visible spectrum; and those for $m = 3, 4,$ and 5, the Paschen, Brackett, and Pfund series, lie in the infrared.

Bohr started with a model suggested by the New Zealand-born British physicist Ernest Rutherford. The model was based on the experiments of Hans Geiger and Ernest Marsden, who in 1909 passed a stream of massive, fast-moving alpha particles through a thin sheet of gold foil. The alpha particles were emitted by a radioactive material and had enough energy to penetrate an atom; although most passed right through the gold foil, some were deflected backward in a way that indicated that the scattering was produced by a Coulomb force. Because the alpha particles are positively charged and the electrons in the atom are negatively charged, it followed that there must be a large positive charge inside the atom to create the Coulomb force by interacting with the alpha particles.

Rutherford concluded that the atom has a massive, charged nucleus. In Rutherford's model, the atom resembles a miniature solar system with the nucleus acting as the Sun and the electrons as the circulating planets. Bohr made three assumptions. First, he postulated that, in contrast to classical mechanics, where an infinite number of orbits is possible, an electron can be in only one of a discrete set of orbits, which he termed stationary states. (Stationary states are also called energy levels. The lowest energy level of a system is called its ground state; higher energy levels are called excited states.) Second, he postulated that the only orbits allowed are those for which the angular momentum of the electron is a whole number n times $\hbar$ ($\hbar = h/2\pi$). Third, Bohr assumed that Newton's laws of motion, so successful in calculating the paths of the planets around the Sun, also applied to electrons orbiting the nucleus. The force on the electron (the analogue of the gravitational force between the Sun and a planet) is the electrostatic attraction between the positively charged nucleus and the negatively charged electron. With these simple assumptions, he showed that the energy of the orbit has the form

$$E_n = -\frac{E_0}{n^2},$$

where E_0 is a constant that may be expressed by a combination of the known constants e, m_e, and $\hbar$. While in a stationary state, the atom does not give off energy as light; however, when an electron makes a transition from a state with energy E_n to one with lower energy E_m, a quantum of energy is radiated with frequency ν, given by the equation

$$h\nu = E_n - E_m.$$

Inserting the expression for E_n into this equation and using the relation $\lambda\nu = c$, where c is the speed of light, Bohr derived the formula for the wavelengths of the lines in the hydrogen spectrum, with the correct value of the Rydberg constant.

Bohr's theory was a brilliant step forward. Its two most important features have survived in present-day quantum mechanics. They are (1) the existence of stationary, nonradiating states and (2) the relationship of radiation frequency to the energy difference between the initial and final states in a transition. Prior to Bohr, physicists had thought that the radiation frequency would be the same as the electron's frequency of rotation in an orbit. The Bohr model of the atom, a radical departure from earlier, classical descriptions, was the first that incorporated quantum theory and was the predecessor of wholly quantum-mechanical models.

The first experimental verification of the existence of discrete energy states in atoms was performed in 1914 by the German-born physicists James Franck and Gustav Hertz. Franck and Hertz directed low-energy electrons through a gas enclosed in an electron tube. As the energy of the electrons was slowly increased, a certain critical electron energy was reached at which the electron stream made a change from almost undisturbed passage through the gas to nearly complete stoppage. The gas atoms were able to absorb the energy of the electrons only when it reached a certain critical value, indicating that within the gas atoms themselves the atomic electrons make an abrupt transition to a discrete higher energy level. As long as the bombarding electrons have less than this discrete amount of energy, no transition is possible and no energy is absorbed from the stream of electrons. When they have this precise energy, they lose it all at once in collisions to atomic electrons, which store the energy by being promoted to a higher energy level.

Scattering of X-rays

Soon scientists were faced with the fact that another form of radiation, X-rays, also exhibits both wave and particle properties. Max von Laue of Germany had shown in 1912 that crystals can be used as three-dimensional diffraction gratings for X-rays; his technique constituted the fundamental evidence for the wavelike nature of X-rays. The atoms of a crystal, which are arranged in a regular lattice, scatter the X-rays. For certain directions of scattering, all the crests of the X-rays coincide. (The scattered X-rays are said to be in phase and to give constructive interference.) For these directions, the scattered X-ray beam is very intense. Clearly, this phenomenon demonstrates wave behaviour. In fact, given the interatomic distances in the crystal and the directions of constructive interference, the wavelength of the waves can be calculated.

In 1922 the American physicist Arthur Holly Compton showed that X-rays scatter from electrons as if they are particles. Compton performed a series of experiments on the scattering of monochromatic, high-energy X-rays by graphite. He found that part of the scattered radiation had the same wavelength λ_o as the incident X-rays but that there was an additional component with a longer wavelength λ. To interpret his results, Compton regarded the X-ray photon as a particle that collides and bounces off an electron in the graphite target as though the photon and the electron were a pair of (dissimilar) billiard balls. Application of the laws of conservation of energy and momentum to the collision leads to a specific relation between the amount of energy transferred to the electron and the angle of scattering. For X-rays scattered through an angle θ, the wavelengths λ and λ_o are related by the equation

$$\lambda - \lambda_0 = \frac{h}{m_e c}(1 - \cos\Theta).$$

The experimental correctness of Compton's formula is direct evidence for the corpuscular behaviour of radiation.

Broglie's Wave Hypothesis

Faced with evidence that electromagnetic radiation has both particle and wave characteristics, Louis-Victor de Broglie of France suggested a great unifying hypothesis in 1924. Broglie proposed that matter has wave, as well as particle, properties. He suggested that material particles can behave as waves and that their wavelength λ is related to the linear momentum p of the particle by $\lambda = h/p$.

In 1927 Clinton Davisson and Lester Germer of the United States confirmed Broglie's hypothesis for electrons. Using a crystal of nickel, they diffracted a beam of monoenergetic electrons and showed that the wavelength of the waves is related to the momentum of the electrons by the Broglie equation. Since Davisson and Germer's investigation, similar experiments have been performed with atoms, molecules, neutrons, protons, and many other particles. All behave like waves with the same wavelength-momentum relationship. Objects of everyday experience, however, have a computed wavelength much smaller than that of electrons, so their wave properties have never been detected; familiar objects show only particle behaviour. De Broglie waves play an appreciable role, therefore, only in the realm of subatomic particles.

The complementarity principle builds on Broglie's wave hypothesis in stating that a complete knowledge of phenomena on atomic dimensions requires a description of both wave and particle properties. The principle was announced in 1928 by the Danish physicist Niels Bohr. Depending on the experimental arrangement, the behaviour of such phenomena as light and electrons is sometimes wavelike and sometimes particle-like; i.e., such things have a wave-particle duality. It is impossible to observe both the wave and particle aspects simultaneously. Together, however, they present a fuller description than either of the two taken alone.

In effect, the complementarity principle implies that phenomena on the atomic and subatomic scale are not strictly like large-scale particles or waves (e.g., billiard balls and water waves). Such particle and wave characteristics in the same large-scale phenomenon are incompatible rather than complementary. Knowledge of a small-scale phenomenon, however, is essentially incomplete until both aspects are known.

Basic Concepts and Methods

Bohr's theory, which assumed that electrons moved in circular orbits, was extended by the German physicist Arnold Sommerfeld and others to include elliptic orbits and other refinements. Attempts were made to apply the theory to more complicated systems than the hydrogen atom. However, the ad hoc mixture of classical and quantum ideas made the theory and calculations increasingly unsatisfactory. Then in 12 months starting in July 1925, a period of creativity without parallel in the history of physics, there appeared a series of papers by German scientists that set the subject on a firm conceptual foundation. The papers took two approaches: (1) matrix mechanics, proposed by Werner Heisenberg, Max Born, and Pascual Jordan, and (2) wave mechanics, put forward by Erwin Schrödinger. The protagonists were not always polite to each other. Heisenberg found the physical ideas of Schrödinger's theory "disgusting," and Schrödinger was "discouraged and repelled" by the lack of visualization in Heisenberg's method. However, Schrödinger, not allowing his emotions to interfere with his scientific endeavours, showed that, in spite of apparent dissimilarities, the two theories are equivalent mathematically. The present discussion follows Schrödinger's wave mechanics because it is less abstract and easier to understand than Heisenberg's matrix mechanics.

SCHRÖDINGER'S WAVE MECHANICS

Schrödinger expressed Broglie's hypothesis concerning the wave behaviour of matter in a mathematical form that is adaptable to a variety of physical problems without additional arbitrary assumptions. He was guided by a mathematical formulation of optics, in which the straight-line propagation of light rays can be derived from wave motion when the wavelength is small compared to the dimensions of the apparatus employed. In the same way, Schrödinger set out to find a wave equation for matter that would give particle-like propagation when the wavelength becomes comparatively small. According to classical mechanics, if a particle of mass m_e is subjected to a force such that its potential energy is $V(x, y, z)$ at position x, y, z, then the sum of $V(x, y, z)$ and the kinetic energy $p^2/2m_e$ is equal to a constant, the total energy E of the particle. Thus,

$$\frac{p^2}{2m_e} + V(x,y,z) = E.$$

It is assumed that the particle is bound—i.e., confined by the potential to a certain region in space because its energy E is insufficient for it to escape. Since the potential varies with position, two other quantities do also: the momentum and, hence, by extension from the Broglie relation, the wavelength of the wave. Postulating a wave function $\Psi(x, y, z)$ that varies with position, Schrödinger replaced p in the above energy equation with a differential operator that embodied the Broglie relation. He then showed that Ψ satisfies the partial differential equation

$$-\frac{\hbar^2}{2m_e}\left(\frac{\delta^2\Psi}{\delta x^2} + \frac{\delta^2\Psi}{\delta y^2} + \frac{\delta^2\Psi}{\delta z^2}\right) + V(x,y,z)\Psi = E\Psi.$$

This is the (time-independent) Schrödinger wave equation, which established quantum mechanics in a widely applicable form. The equation has the same central importance to quantum mechanics as Newton's laws of motion have for the large-scale phenomena of classical mechanics. An important advantage of Schrödinger's theory is that no further arbitrary quantum conditions need be postulated. The required quantum results follow from certain reasonable restrictions placed on the wave function—for example, that it should not become infinitely large at large distances from the centre of the potential.

Schrödinger applied his equation to the hydrogen atom, for which the potential function, given by classical electrostatics, is proportional to $-e^2/r$, where $-e$ is the charge on the electron. The nucleus (a proton of charge e) is situated at the origin, and r is the distance from the origin to the position of the electron. Schrödinger solved the equation for this particular potential with straightforward, though not elementary, mathematics, predicting many of the hydrogen atom's properties with remarkable accuracy. Only certain discrete values of E lead to acceptable functions Ψ. These functions are characterized by a trio of integers n, l, m, termed quantum numbers.

Quantum numbers are any of several quantities of integral or half-integral value that identify the state of a physical system such as an atom, a nucleus, or a subatomic particle. They refer generally to properties that are discrete (quantized) and conserved, such as energy, momentum, charge, baryon number, and lepton number.

The principal quantum number for electrons, n, confined in atoms, for example, indicates the energy state and the probability of finding the electrons at various distances from the nucleus. The larger the principal quantum number, which has integral values beginning with one, the greater the energy is and the farther the electron is likely to be from the nucleus.

The values of E depend only on the integers n (1, 2, 3, etc.) and are identical with those given by the Bohr theory. The quantum numbers l and m are related to the angular momentum of the electron; $\sqrt{l(l+1)}\hbar$ is the magnitude of the angular momentum, and $m\hbar$ is its component along some physical direction.

The square of the wave function, Ψ^2, has a physical interpretation. Schrödinger originally supposed that the electron was spread out in space and that its density at point x, y, z was given by the value of Ψ^2 at that point. Almost immediately Born proposed what is now the accepted interpretation—namely, that Ψ^2 gives the probability of finding the electron at x, y, z. The distinction between the two interpretations is important. If Ψ^2 is small at a particular position, the original interpretation implies that a small fraction of an electron will always be detected there. In Born's interpretation, nothing will be detected there most of the time, but, when something is observed, it will be a whole electron. Thus, the concept of the electron as a point particle moving in a well-defined path around the nucleus is replaced in wave mechanics by clouds that describe the probable locations of electrons in different states.

ELECTRON SPIN AND ANTIPARTICLES

In 1928 the English physicist Paul A.M. Dirac produced a wave equation for the electron that combined relativity with quantum mechanics. Schrödinger's wave equation does not satisfy the requirements of the special theory of relativity because it is based on a nonrelativistic expression for the kinetic energy ($p^2/2m_e$). Dirac showed that an electron has an additional quantum number m_s. Unlike the first three quantum numbers, m_s is not a whole integer and can have only the values + ½ and - ½. It corresponds to an additional form of angular momentum ascribed to a spinning motion. (The angular momentum mentioned above is due to the orbital motion of the electron, not its spin.) The concept of spin angular momentum was introduced in 1925 by Samuel A. Goudsmit and George E. Uhlenbeck, two graduate students at the University of Leiden, Neth., to explain the magnetic moment measurements made by Otto Stern and Walther Gerlach of Germany several years earlier. The magnetic moment of a particle is closely related to its angular momentum; if the angular momentum is zero, so is the magnetic moment.

In Stern and Gerlach's experiment, a beam of neutral silver atoms was directed through a set of aligned slits, then through a nonuniform (nonhomogeneous) magnetic field, and onto a cold glass plate. An electrically neutral silver atom is actually an atomic magnet: the spin of an unpaired electron causes the atom to have a north and south pole like a tiny compass needle. In a uniform

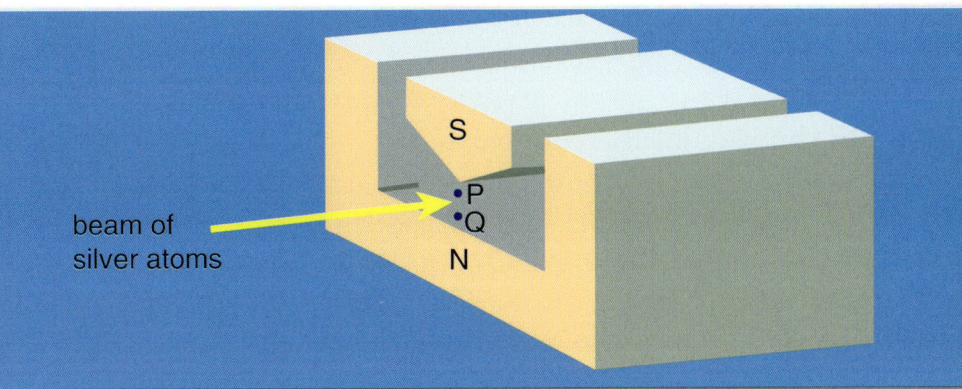

A beam of silver atoms is passed between the north (N) and south (S) poles of a magnet. The poles are shaped so that the magnetic field varies greatly in strength over a very small distance. The knife-edge of S results in a much stronger magnetic field at point P than at point Q. Copyright Encyclopædia Britannica; rendering for this edition by Rosen Educational Services

magnetic field, the atomic magnet, or magnetic dipole, only precesses as the atom moves in the external magnetic field. In a nonuniform magnetic field, the forces on the two poles are not equal, and the silver atom itself is deflected by a slight resultant force, the magnitude and direction of which vary in relation to the orientation of the dipole in the nonuniform field. A beam of neutral silver atoms directed through the apparatus in the absence of the nonuniform magnetic field produces a thin line, in the shape of the slit, on the plate. When the nonuniform magnetic field is applied, the thin line splits lengthwise into two distinct traces, corresponding to just two opposite orientations in space of the silver atoms. If the silver atoms were oriented randomly in space, the trace on the plate would have broadened into a wide area, corresponding to numerous different deflections of the silver atoms. Stern and Gerlach had observed a magnetic moment for electrons in silver atoms, which were known to have zero orbital angular momentum. Goudsmit and Uhlenbeck proposed that the observed magnetic moment was attributable to spin angular momentum.

The electron-spin hypothesis not only provided an explanation for the observed magnetic moment but also accounted for many other effects in atomic spectroscopy, including doublet lines in alkali spectra and fine structure (close doublets and triplets) in the hydrogen spectrum.

Fine structure is produced when an atom emits light in making the transition from one energy state to another. The split lines, which are called the fine structure of the main lines, arise from the interaction of the orbital motion of an electron with the quantum mechanical "spin" of that electron. An electron can be thought of as an electrically charged spinning top, and hence it behaves as a tiny bar magnet. The spinning electron interacts with the magnetic field produced by the electron's rotation about the atomic nucleus to generate the fine structure.

The amount of splitting is characterized by a dimensionless constant called the fine-structure constant. This constant is given by the equation $\alpha = ke^2/hc$, where k is Coulomb's constant, e is the charge of the electron, h is Planck's constant, and c is the speed of light. The value of the constant α is $7.29735254 \times 10^{-3}$, which is nearly equal to $1/137$.

In the atoms of alkali metals such as sodium and potassium, there are two components of fine structure (called doublets), while in atoms of alkaline earths there are three components (triplets). This arises because the atoms of alkali metals have only one electron outside a closed core, or shell, of electrons, while the atoms of alkaline earths have two such electrons. Doublet separation for corresponding lines increases with atomic number; thus, with lithium (atomic number 3), a doublet may not be resolved by an ordinary spectroscope, whereas with rubidium (atomic number 37), a doublet may be widely separated.

There is a further splitting of a spectral line into a number of components called hyperfine structure. The splitting is caused by nuclear effects and cannot be observed in an ordinary spectroscope without the aid of an optical device called an interferometer. In fine structure, line splitting is the result of energy changes produced by electron spin–orbit coupling (i.e., interaction of forces from orbital and spin motion of electrons); but in hyperfine structure, line splitting is attributed to the fact that in addition to electron spin in an atom, the atomic nucleus itself spins about its own axis. Energy states of the atom will be split into levels corresponding to slightly different energies. Each of these energy levels may be assigned a quantum number, and they are then called quantized levels. Thus, when the atoms of an element radiate energy, transitions are made between these quantized energy levels, giving rise to hyperfine structure.

The spin quantum number is zero for nuclei of even atomic number and even mass number, and therefore no hyperfine structure is found in their spectral lines. The spectra of other nuclei do exhibit hyperfine structure. By observing hyperfine structure, it is possible to calculate nuclear spin.

A similar effect of line splitting is caused by mass differences (isotopes) of atoms in an element and is called isotope structure, or isotope shift. These spectral lines are sometimes referred to as hyperfine structure but may be observed in an element with spin-zero isotopes (even atomic and mass numbers). Isotope structure is seldom observed without true hyperfine structure accompanying it.

The electron-spin hypothesis also explained the Zeeman effect, the splitting of a spectral line into two or more components of slightly different frequency when the light source is placed in a magnetic field. It was first observed in 1896 by the Dutch physicist Pieter Zeeman as a broadening of the yellow D-lines of sodium in a flame held between strong magnetic poles. Later the broadening was found to be a distinct splitting of spectral lines into as many as 15 components.

Zeeman's discovery earned him the 1902 Nobel Prize for Physics, which he shared with a former teacher, Hendrik Antoon Lorentz, another Dutch physicist. Lorentz, who had earlier developed a theory concerning the effect of magnetism on light, hypothesized that the oscillations of electrons inside an atom produce light and that a magnetic field would affect the oscillations and thereby the frequency of the light emitted. This theory was confirmed by Zeeman's research and later modified by quantum mechanics, according to which spectral lines of light are emitted when electrons change from one discrete energy level to another. Each of the levels, characterized by an angular momentum , is split in a magnetic field into substates of equal energy. These substates of energy are revealed by the resulting patterns of spectral line components.

The Zeeman effect has helped physicists determine the energy levels in atoms and identify them in terms of angular momenta. It also provides an effective means of studying atomic nuclei and such phenomena as electron paramagnetic resonance. In astronomy, the Zeeman effect is used in measuring the magnetic field of the Sun and of other stars.

The electric analogue of the Zeeman effect is the Stark effect, which was discovered by a German physicist, Johannes Stark (1913). Earlier experimenters had failed to maintain a strong electric field in conventional spectroscopic light sources because of the high electrical conductivity of luminous gases or vapours. Stark observed the hydrogen spectrum emitted just behind the perforated cathode in a positive-ray tube. With a second charged electrode parallel and close to this cathode, he was able to produce a strong electric field in a space of a few millimetres. At electric field intensities of 100,000 volts per centimetre, Stark observed with a spectroscope that the characteristic spectral lines, called Balmer lines, of hydrogen were split into a number of symmetrically spaced components, some of which were linearly polarized (vibrating in one plane) with the electric vector parallel to the lines of force, the remainder being polarized perpendicular to the direction of the field except when viewed along the field. This transverse Stark effect resembles in some respects the transverse Zeeman effect, but, because of its complexity, the Stark effect has relatively less value in the analysis of complicated spectra or of atomic structure. Historically, the satisfactory explanation of the Stark effect (1916) was one of the great triumphs of early quantum mechanics.

The Dirac equation also predicted additional states of the electron that had not yet been observed. Experimental confirmation was provided in 1932 by the discovery of the positron by the American physicist Carl David Anderson. Every particle described by the Dirac equation has to have a corresponding antiparticle, which differs only in charge. The positron is just such an antiparticle of the negatively charged electron, having the same mass as the latter but a positive charge.

IDENTICAL PARTICLES AND MULTIELECTRON ATOMS

Because electrons are identical to (i.e., indistinguishable from) each other, the wave function of an atom with more than one electron must satisfy special conditions. The problem of identical particles does not arise in classical physics, where the objects are large-scale and can always be distinguished, at least in principle. There is no way, however, to differentiate two electrons in the same atom, and the form of the wave function must reflect this fact. The overall wave function Ψ of a system of identical particles depends on the coordinates of all the particles. If the coordinates of two of the particles are interchanged, the wave function must remain unaltered or, at most, undergo a change of sign; the change of sign is permitted because it is Ψ^2 that occurs in the physical interpretation of the wave function. If the sign of Ψ remains unchanged, the wave function is said to be symmetric with respect to interchange; if the sign changes, the function is antisymmetric.

The symmetry of the wave function for identical particles is closely related to the spin of the particles. In quantum field theory, it can be shown that particles with half-integral spin ($1/2$, $3/2$, etc.) have antisymmetric wave functions. They are called fermions after the Italian-born physicist Enrico Fermi. Examples of fermions are electrons, protons, and neutrons, all of which have spin $1/2$. Particles with zero or integral spin (e.g., mesons, photons) have symmetric wave functions and are called bosons after the Indian mathematician and physicist Satyendra Nath Bose, who first applied the ideas of symmetry to photons in 1924–25.

The requirement of antisymmetric wave functions for fermions leads to a fundamental result, known as the exclusion principle, first proposed in 1925 by the Austrian physicist Wolfgang Pauli. The exclusion principle states that two fermions in the same system cannot be in the same quantum state. If they were, interchanging the two sets of coordinates would not change the wave function at all, which contradicts the result that the wave function must change sign. Thus, two electrons in the same atom cannot have an identical set of values for the four quantum numbers n, l, m, m_s. The exclusion principle forms the basis of many properties of matter, including the periodic classification of the elements, the nature of chemical bonds, and the behaviour of electrons in solids; the last determines in turn whether a solid is a metal, insulator, or semiconductor.

For example, the Pauli exclusion principle leads to the simplified description of the structure of atoms that was first proposed by the physicists J. Hans D. Jensen and Maria Goeppert Mayer working independently in 1949. In this model, electrons in atoms are thought of as occupying diffuse shells in the space surrounding a dense, positively charged nucleus. The first shell is closest to the nucleus. The others extend outward from the nucleus and overlap one another. The shells are sometimes designated by capital letters beginning with K for the first shell, L for the second, M for the third, and so forth through Q for the seventh shell. The maximum number of electrons that can occupy shells one through seven are, in sequence, 2, 8, 18, 32, 50, 72, 98. The lightest element, hydrogen, has one electron in the first shell. The heaviest elements in their normal states have only the first four shells fully occupied with electrons and the next three shells partially occupied. In terms of a more refined, quantum-mechanical model, the K–Q shells are subdivided into a set of orbitals, each of which can be occupied by no more than a pair of electrons.

Atomic orbitals are commonly designated by a combination of numerals and letters that represent specific properties of the electrons associated with the orbitals — for example, $1s$, $2p$, $3d$, $4f$. The numerals, called principal quantum numbers, indicate energy levels as well as relative distance from the nucleus. A $1s$ electron occupies the energy level nearest the nucleus. A $2s$ electron, less strongly bound, spends most of its time farther away from the nucleus. The letters, s, p, d, and f designate the shape of the orbital. (The shape is a consequence of the magnitude of the electron's angular momentum, resulting from its angular motion.) An s orbital is spherical with its centre at the nucleus. Thus a 1s electron is almost entirely confined to a spherical region close to the nucleus; a 2s electron is restricted to a somewhat larger sphere. A p orbital has the approximate shape of a pair of lobes on opposite sides of the nucleus, or a somewhat dumbbell shape. An electron in a p orbital has equal probability of being in either half. The shapes of the other orbitals are more complicated. The letters s, p, d, f originally were used to classify spectra descriptively into series called sharp, principal, diffuse, and fundamental, before the relation between spectra and atomic electron configuration was known.

No p orbitals exist in the first energy level, but there is a set of three in each of the higher levels. These triplets are oriented in space as if they were on three axes at right angles to each other and may be distinguished by subscripts, for example, $2p_x$, $2p_y$, $2p_z$. In all but the first two principal levels, there is a set of five d orbitals and, in all but the first three principal levels, a set of seven f orbitals, all with complicated orientations.

The electronic configuration of an atom in the quantum-mechanical model is stated by listing the occupied orbitals, in order of filling, with the number of electrons in each orbital indicated by superscript. In this notation, the electronic configuration of sodium would be $1s^2 2s^2 2p^6 3s^1$, distributed in the orbitals as 2-8-1. Often, a shorthand method is used that lists only those electrons in excess of the noble gas configuration immediately preceding the atom in the periodic table. For example, sodium has one $3s$ electron in excess of the noble gas neon (chemical symbol Ne, atomic number 10), and so its shorthand notation is [Ne]$3s^1$.

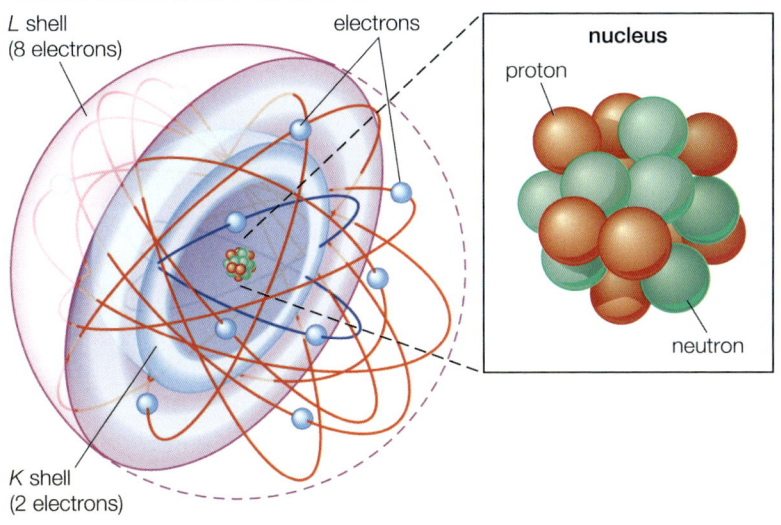

In the shell atomic model, electrons occupy different energy levels, or shells. The K and L shells are shown for a neon atom. Encyclopædia Britannica, Inc.

The chemical properties of atoms are explained in terms of how the shells are occupied with electrons. For example, helium (atomic number 2) has a full first shell; neon (atomic number 10), with eight electrons in its outermost shell, has a full first and second shell.

Other atoms that have eight electrons in their outermost shell, even though it is not full, chemically resemble helium and neon in their relative stability and inactivity.

The Schrödinger equation cannot be solved precisely for atoms with more than one electron. The principles of the calculation are well understood, but the problems are complicated by the number of particles and the variety of forces involved. The forces include the electrostatic forces between the nucleus and the electrons and between the electrons themselves, as well as weaker magnetic forces arising from the spin and orbital motions of the electrons. Despite these difficulties, approximation methods introduced by the English physicist Douglas

R. Hartree and others in the 1920s have achieved considerable success. Such schemes start by assuming that each electron moves independently in an average electric field because of the nucleus and the other electrons—i.e., correlations between the positions of the electrons are ignored. Each electron has its own wave function, called an orbital. The overall wave function for all the electrons in the atom satisfies the exclusion principle. Corrections to the calculated energies are then made, which depend on the strengths of the electron-electron correlations and the magnetic forces.

TIME-DEPENDENT SCHRÖDINGER EQUATION

At the same time that Schrödinger proposed his time-independent equation to describe the stationary states, he also proposed a time-dependent equation to describe how a system changes from one state to another. By replacing the energy E in Schrödinger's equation with a time-derivative operator, he generalized his wave equation to determine the time variation of the wave function as well as its spatial variation. The time-dependent Schrödinger equation reads

$$-\frac{\hbar^2}{2m_e}\left(\frac{\delta^2 \Psi}{\delta x^2} + \frac{\delta^2 \Psi}{\delta y^2} + \frac{\delta^2 \Psi}{\delta z^2}\right) + V(x,y,z)\Psi = i\hbar \frac{\delta \Psi}{\delta t}.$$

The quantity i is the square root of -1. The function Ψ varies with time t as well as with position x, y, z. For a system with constant energy, E, Ψ has the form

$$\Psi(x,y,z,t) = \Psi(x,y,z)\exp{-\frac{iEt}{\hbar}},$$

where exp stands for the exponential function, and the

time-dependent Schrödinger equation reduces to the time-independent form.

The probability of a transition between one atomic stationary state and some other state can be calculated with the aid of the time-dependent Schrödinger equation. For example, an atom may change spontaneously from one state to another state with less energy, emitting the difference in energy as a photon with a frequency given by the Bohr relation. If electromagnetic radiation is applied to a set of atoms and if the frequency of the radiation matches the energy difference between two stationary states, transitions can be stimulated. In a stimulated transition, the energy of the atom may increase—i.e., the atom may absorb a photon from the radiation—or the energy of the atom may decrease, with the emission of a photon, which adds to the energy of the radiation. Such stimulated emission processes form the basic mechanism for the operation of lasers. The probability of a transition from one state to another depends on the values of the l, m, m_s quantum numbers of the initial and final states. For most values, the transition probability is effectively zero. However, for certain changes in the quantum numbers, summarized as selection rules, there is a finite probability. For example, according to one important selection rule, the l value changes by unity because photons have a spin of 1. The selection rules for radiation relate to the angular momentum properties of the stationary states. The absorbed or emitted photon has its own angular momentum, and the selection rules reflect the conservation of angular momentum between the atoms and the radiation.

A similar process happens in the Auger effect, a spontaneous process in which an atom with an electron vacancy in the innermost (K) shell readjusts itself to a more stable state by ejecting one or more electrons instead of radiating a single X-ray photon. This internal photoelectric process is named for the French physicist Pierre-Victor Auger, who discovered it in 1925.

All atoms consist of a nucleus and concentric shells of electrons. If an electron in one of the inner shells is removed by electron bombardment, absorption into the nucleus, or in some other way, an electron from another shell will jump into the vacancy, releasing energy that is promptly dissipated either by producing an X-ray or through the Auger effect. In the Auger effect, the available energy expels an electron from one of the shells with the result that the residual atom then has two electron vacancies. The process may be repeated as the new vacancies are filled, otherwise X-radiation will be emitted. The probability that an Auger electron will be emitted is called the Auger yield for that shell. The Auger yield decreases with atomic number (the number of protons in the nucleus), and at atomic number 30 (zinc) the probabilities of the emission of X-rays from the innermost shell and of the emission of Auger electrons is about equal. The Auger effect is useful in studying the properties of elements and compounds, nuclei, and subatomic particles called muons.

TUNNELING

The phenomenon of tunneling, also called barrier penetration, which has no counterpart in classical physics, is an important consequence of quantum mechanics. Consider a particle with energy E in the inner region of a one-dimensional potential well $V(x)$. (A potential well is a potential that has a lower value in a certain region of space than in the neighbouring regions.) In classical mechanics, if $E < V_o$ (the maximum height of the potential barrier), the particle remains in the well forever; if $E > V_o$, the

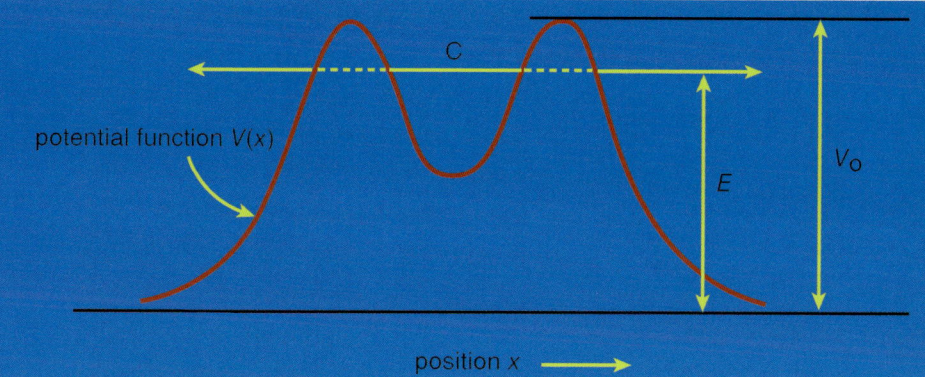

The phenomenon of tunneling. Classically, a particle is bound in the central region C if its energy E is less than V_o, but in quantum theory the particle may tunnel through the potential barrier and escape. Copyright Encyclopædia Britannica; rendering for this edition by Rosen Educational Services

particle escapes. In quantum mechanics, the situation is not so simple. The particle can escape even if its energy E is below the height of the barrier V_o, although the probability of escape is small unless E is close to V_o. In that case, the particle may tunnel through the potential barrier and emerge with the same energy E.

The phenomenon of tunneling has many important applications. For example, it describes a type of radioactive decay in which a nucleus emits an alpha particle (a helium nucleus). According to the quantum explanation given independently by George Gamow and by Ronald W. Gurney and Edward Condon in 1928, the alpha particle is confined before the decay by a potential well. For a given nuclear species, it is possible to measure the energy E of the emitted alpha particle and the average lifetime τ of the nucleus before decay. The lifetime of the nucleus is a measure of the probability of tunneling through the barrier—the shorter the lifetime, the higher the probability. With plausible assumptions about the general form of the potential function, it is possible to calculate a relationship between τ and E that is applicable to all alpha emitters. This theory, which is borne out by experiment, shows that the probability of tunneling, and hence the value of τ, is extremely sensitive to the value of E. For all known alpha-particle emitters, the value of E varies from about 2 to 8 million electron volts, or MeV (1 MeV = 10^6 electron volts). Thus, the value of E varies only by a factor of 4, whereas the range of τ is from about 10^{11} years down to about 10^{-6} second, a factor of 10^{24}. It would be difficult to account for this sensitivity of τ to the value of E by any theory other than quantum mechanical tunneling.

AXIOMATIC APPROACH

Although the two Schrödinger equations form an important part of quantum mechanics, it is possible to present the subject in a more general way. Dirac gave an elegant exposition of an axiomatic approach based on observables and states in a classic textbook entitled *The Principles of Quantum Mechanics*. (The book, published in 1930, is still in print.) An observable is anything that can be measured—energy, position, a component of angular momentum, and so forth. Every observable has a set of states, each state being represented by an algebraic function. With each state is associated a number that gives the result of a measurement of the observable. Consider an observable with N states, denoted by $\psi_1, \psi_2, \ldots, \psi_N$, and corresponding measurement values $a_1, a_2, \ldots, a_N$. A physical system—e.g., an atom in a particular state—is represented by a wave function Ψ, which can be expressed as a linear combination, or mixture, of the states of the observable. Thus, the Ψ may be written as

$$\Psi = c_1\psi_1 + c_2\psi_2 + \ldots + c_N\psi_N. \tag{10}$$

For a given Ψ, the quantities c_1, c_2, etc., are a set of numbers that can be calculated. In general, the numbers are complex, but, in the present discussion, they are assumed to be real numbers.

The theory postulates, first, that the result of a measurement must be an a-value—i.e., a_1, a_2, or a_3, etc. No other value is possible. Second, before the measurement is made, the probability of obtaining the value a_1 is c_1^2, and that of obtaining the value a_2 is c_2^2, and so on. If the value obtained is, say, a_5, the theory asserts that after the measurement the state of the system is no longer the original Ψ but has changed to ψ_5, the state corresponding to a_5.

A number of consequences follow from these assertions. First, the result of a measurement cannot be predicted with certainty. Only the probability of a particular result can be predicted, even though the initial state (represented by the function Ψ) is known exactly. Second, identical measurements made on a large number of identical systems, all in the identical state Ψ, will produce different values for the measurements. This is, of course, quite contrary to classical physics and common sense, which say that the same measurement on the same object in the same state must produce the same result. Moreover, according to the theory, not only does the act of measurement change the state of the system, but it does so in an indeterminate way. Sometimes it changes the state to ψ_1, sometimes to ψ_2, and so forth.

There is an important exception to the above statements. Suppose that, before the measurement is made, the state Ψ happens to be one of the ψs, say $\Psi = \psi_3$. Then $c_3 = 1$ and all the other cs are zero. This means that, before the measurement is made, the probability of obtaining the value a_3 is unity and the probability of obtaining any other value of a is zero. In other words, in this particular case, the result of the measurement can be predicted with certainty. Moreover, after the measurement is made, the state will be ψ_3, the same as it was before. Thus, in this particular case, measurement does not disturb the system. Whatever the initial state of the system, two measurements made in rapid succession (so that the change in the wave function given by the time-dependent Schrödinger equation is negligible) produce the same result.

The value of one observable can be determined by a single measurement. The value of two observables for a given system may be known at the same time, provided that the two observables have the same set of state functions $\psi_1, \psi_2, \ldots, \psi_N$. In this case, measuring the first observable results in a state function that is one of the ψs. Because this is also a state function of the second observable, the result of measuring the latter can be predicted with certainty. Thus the values of both observables are known. (Although the ψs are the same for the two observables, the two sets of a values are, in general, different.) The two observables can be measured repeatedly in any sequence. After the first measurement, none of the measurements disturbs the system, and a unique pair of values for the two observables is obtained.

INCOMPATIBLE OBSERVABLES

The measurement of two observables with different sets of state functions is a quite different situation. Measurement of one observable gives a certain result. The state function after the measurement is, as always, one of the states of that observable; however, it is not a state function for the second observable. Measuring the second observable disturbs the system, and the state of the system is no longer one of the states of the first observable. In general, measuring the first observable again does not produce the same result as the first time. To sum up, both quantities cannot be known at the same time, and the two observables are said to be incompatible.

A specific example of this behaviour is the measurement of the component of angular momentum along two mutually perpendicular directions. The Stern-Gerlach experiment mentioned above involved measuring the angular momentum of a silver atom in the ground state. In reconstructing this experiment, a beam of silver atoms is passed between the poles of a magnet. The poles are shaped so that the magnetic field varies greatly in strength over a very small distance. The apparatus determines the m_s quantum number, which can be + ½ or - ½. No other values are obtained. Thus in this case the observable has only two states—i.e., $N = 2$. The inhomogeneous magnetic field produces a force on the silver atoms in a direction that depends on the spin state of the atoms. Consider a beam of silver atoms that is passed through magnet A. The atoms in the state with m_s = + ½ are deflected upward and emerge as beam 1, while those with m_s = - ½ are deflected downward and emerge as beam 2. If the direction of the magnetic field is the x-axis, the apparatus measures S_x, which is the x-component of spin angular momentum. The atoms in beam 1 have $S_x = +\hbar/2$ while those in beam 2 have $S_x = -\hbar/2$. In a classical picture, these two states represent atoms spinning about the direction of the x-axis with opposite senses of rotation.

The y-component of spin angular momentum S_y also can have only the values $+\hbar/2$ and $-\hbar/2$; however, the two states of S_y are not the same as for S_x. In fact, each of the states of S_x is an equal mixture of the states for S_y, and conversely. Again, the two S_y states may be pictured as representing atoms with opposite senses of rotation about the y-axis. These classical pictures of quantum states are helpful, but only up to a certain point. For example, quantum

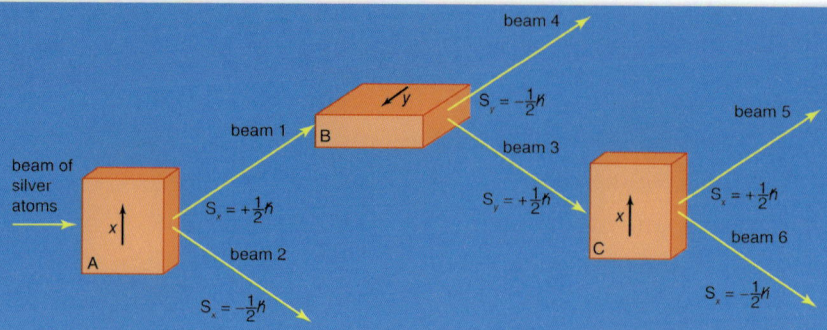

Measurements of the x and y components of angular momentum for silver atoms, S, in the ground state. A, B, and C are magnets with inhomogeneous magnetic fields. The arrows show the average direction of each magnetic field. Copyright Encyclopædia Britannica; rendering for this edition by Rosen Educational Services

theory says that each of the states corresponding to spin about the *x*-axis is a superposition of the two states with spin about the *y*-axis. There is no way to visualize this; it has absolutely no classical counterpart. One simply has to accept the result as a consequence of the axioms of the theory. Suppose that the atoms in beam 1 are passed into a second magnet B, which has a magnetic field along the *y*-axis perpendicular to *x*. The atoms emerge from B and go in equal numbers through its two output channels. Classical theory says that the two magnets together have measured both the *x*- and *y*-components of spin angular momentum and that the atoms in beam 3 have $S_x = +\hbar/2$, $S_y = +\hbar/2$, while those in beam 4 have $S_x = +\hbar/2$, $S_y = -\hbar/2$. However, classical theory is wrong, because if beam 3 is put through still another magnet C, with its magnetic field along *x*, the atoms divide equally into beams 5 and 6 instead of emerging as a single beam 5 (as they would if they had $S_x = +\hbar/2$). Thus, the correct statement is that the beam entering B has $S_x = +\hbar/2$ and is composed of an equal mixture of the states $S_y = +\hbar/2$ and $S_y = -\hbar/2$—i.e., the *x*-component of angular momentum is known but the *y*-component is not. Correspondingly, beam 3 leaving B has

$S_y = +\hbar/2$ and is an equal mixture of the states $S_x = +\hbar/2$ and $S_x = -\hbar/2$; the y-component of angular momentum is known but the x-component is not. The information about S_x is lost because of the disturbance caused by magnet B in the measurement of S_y.

HEISENBERG UNCERTAINTY PRINCIPLE

The observables discussed so far have had discrete sets of experimental values. For example, the values of the energy of a bound system are always discrete, and angular momentum components have values that take the form $m\hbar$, where m is either an integer or a half-integer, positive or negative. On the other hand, the position of a particle or the linear momentum of a free particle can take continuous values in both quantum and classical theory. The mathematics of observables with a continuous spectrum of measured values is somewhat more complicated than for the discrete case but presents no problems of principle. An observable with a continuous spectrum of measured values has an infinite number of state functions. The state function Ψ of the system is still regarded as a combination of the state functions of the observable, but the sum in equation (10) must be replaced by an integral.

Measurements can be made of position x of a particle and the x-component of its linear momentum, denoted by p_x. These two observables are incompatible because they have different state functions. The phenomenon of diffraction noted above illustrates the impossibility of measuring position and momentum simultaneously and precisely. If a parallel monochromatic light beam passes through a slit, its intensity varies with direction. The light has zero intensity in certain directions. Wave theory shows that the first zero occurs at an angle θ_o, given by

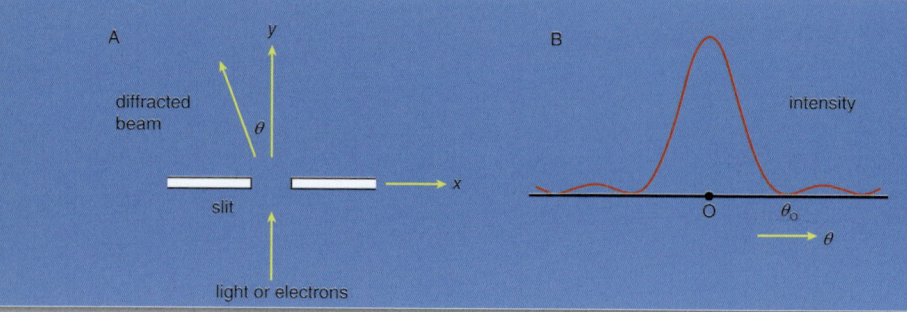

(A) Parallel monochromatic light incident normally on a slit, (B) variation in the intensity of the light with direction after it has passed through the slit. If the experiment is repeated with electrons instead of light, the same diagram would represent the variation in the intensity (i.e., relative number) of the electrons. Copyright Encyclopædia Britannica; rendering for this edition by Rosen Educational Services

$\sin \theta_o = \lambda/b$, where λ is the wavelength of the light and b is the width of the slit. If the width of the slit is reduced, θ_o increases—i.e., the diffracted light is more spread out. Thus, θ_o measures the spread of the beam.

The experiment can be repeated with a stream of electrons instead of a beam of light. According to Broglie, electrons have wavelike properties; therefore, the beam of electrons emerging from the slit should widen and spread out like a beam of light waves. This has been observed in experiments. If the electrons have velocity u in the forward direction (i.e., the y-direction), their (linear) momentum is $p = m_e u$. Consider p_x, the component of momentum in the x-direction. After the electrons have passed through the aperture, the spread in their directions results in an uncertainty in p_x by an amount

$$\Delta p_x \gg p \sin \Theta_o = p\frac{\lambda}{b}, \qquad (11)$$

where λ is the wavelength of the electrons and, according to the Broglie formula, equals h/p. Thus, $\Delta p_x \approx h/b$. Exactly where an electron passed through the slit is unknown; it is only certain that an electron went through somewhere. Therefore, immediately after an electron goes through, the uncertainty in its x-position is $\Delta x \approx b/2$. Thus the product of the uncertainties is of the order of $\hbar$. More exact analysis shows that the product has a lower limit, given by

$$\Delta x \Delta p_x \geq \frac{\hbar}{2}. \tag{12}$$

This is the well-known Heisenberg uncertainty principle for position and momentum. It states that there is a limit to the precision with which the position and the momentum of an object can be measured at the same time. Depending on the experimental conditions, either quantity can be measured as precisely as desired (at least in principle), but the more precisely one of the quantities is measured, the less precisely the other is known. The very concepts of exact position and exact momentum together, in fact, have no meaning in nature.

The uncertainty principle is significant only on the atomic scale because of the small value of $\hbar$ in everyday units. If the position of a macroscopic object with a mass of, say one gram is measured with a precision of 10^{-6} metre, the uncertainty principle states that its velocity cannot be measured to better than about 10^{-25} metre per second. Such a limitation is hardly worrisome. However, if an electron is located in an atom about 10^{-10} metre across, the principle gives a minimum uncertainty in the velocity of about 10^6 metre per second.

The above reasoning leading to the uncertainty principle is based on the wave-particle duality of the electron. When Heisenberg first propounded the principle in 1927 his reasoning was based, however, on the wave-particle duality of the photon. He considered the process of measuring the position of an electron by observing it in a microscope. Diffraction effects due to the wave nature of light result in a blurring of the image; the resulting uncertainty in the position of the electron is approximately equal to the wavelength of the light. To reduce this uncertainty, it is necessary to use light of shorter wavelength—e.g., gamma rays. However, in producing an image of the electron, the gamma-ray photon bounces off the electron, giving the Compton effect. As a result of the collision, the electron recoils in a statistically random way. The resulting uncertainty in the momentum of the electron is proportional to the momentum of the photon, which is inversely proportional to the wavelength of the photon. So it is again the case that increased precision in knowledge of the position of the electron is gained only at the expense of decreased precision in knowledge of its momentum. A detailed calculation of the process yields the same result as before (equation [12]). Heisenberg's reasoning brings out clearly the fact that the smaller the particle being observed, the more significant is the uncertainty principle. When a large body is observed, photons still bounce off it and change its momentum, but, considered as a fraction of the initial momentum of the body, the change is insignificant.

The Schrödinger and Dirac theories give a precise value for the energy of each stationary state, but in reality the states do not have a precise energy. The only exception is in the ground (lowest energy) state. Instead, the energies of the states are spread over a small range. The spread arises from the fact that, because the electron can make a transition to another state, the initial state has a finite lifetime. The transition is a random process, and so different atoms in the same state have different lifetimes. If the mean lifetime is denoted as τ, the theory shows that the energy of the initial state has a spread of energy ΔE,

given by
$$t\Delta E \approx \hbar. \qquad (13)$$

This energy spread is manifested in a spread in the frequencies of emitted radiation. Therefore, the spectral lines are not infinitely sharp. (Some experimental factors can also broaden a line, but their effects can be reduced; however, the present effect, known as natural broadening, is fundamental and cannot be reduced.) Equation (13) is another type of Heisenberg uncertainty relation; generally, if a measurement with duration τ is made of the energy in a system, the measurement disturbs the system, causing the energy to be uncertain by an amount ΔE, the magnitude of which is given by the above equation.

QUANTUM ELECTRODYNAMICS

The application of quantum theory to the interaction between electrons and radiation requires a quantum treatment of Maxwell's field equations, which are the foundations of electromagnetism, and the relativistic theory of the electron formulated by Dirac. The resulting quantum field theory is known as quantum electrodynamics, or QED.

QED accounts for the behaviour and interactions of electrons, positrons, and photons. It deals with processes involving the creation of material particles from electromagnetic energy and with the converse processes in which a material particle and its antiparticle annihilate each other and produce energy. Initially the theory was beset with formidable mathematical difficulties, because the calculated values of quantities such as the charge and mass of the electron proved to be infinite. However, an ingenious set of techniques developed (in the late 1940s) by Hans Bethe, Julian S. Schwinger, Tomonaga Shin'ichirō, Richard

P. Feynman, and others dealt systematically with the infinities to obtain finite values of the physical quantities. Their method is known as renormalization. The theory has provided some remarkably accurate predictions.

According to the Dirac theory, two particular states in hydrogen with different quantum numbers have the same energy. QED, however, predicts a small difference in their energies; the difference may be determined by measuring the frequency of the electromagnetic radiation that produces transitions between the two states. This effect was first measured by Willis E. Lamb, Jr., and Robert Retherford in 1947. Its physical origin lies in the interaction of the electron with the random fluctuations in the surrounding electromagnetic field. These fluctuations, which exist even in the absence of an applied field, are a quantum phenomenon. The accuracy of experiment and theory in this area may be gauged by two recent values for the separation of the two states, expressed in terms of the frequency of the radiation that produces the transitions:

```
experiment (1982)    1,057,858 ± 2 kilohertz
theory (1975)        1,057,864 ±14 kilohertz.
```

An even more spectacular example of the success of QED is provided by the value for μ_e, the magnetic dipole moment of the free electron. Because the electron is spinning and has electric charge, it behaves like a tiny magnet, the strength of which is expressed by the value of μ_e. According to the Dirac theory, μ_e is exactly equal to $\mu_B = e\hbar/2m_e$, a quantity known as the Bohr magneton; however, QED predicts that $\mu_e = (1 + a)\mu_B$, where a is a small number, approximately $1/860$. Again, the physical origin of the QED correction is the interaction of the electron with random oscillations in the surrounding electromagnetic field. The best experimental determination of μ_e involves measuring not the quantity itself but the small correction term $\mu_e - \mu_B$. This greatly enhances the sensitivity of the experiment. The most recent results for the value of a are

experiment (1984) $(115{,}965{,}219 \pm 1) \times 10^{-11}$

theory (1986) $(115{,}965{,}227 \pm 10) \times 10^{-11}$.

Since a itself represents a small correction term, the magnetic dipole moment of the electron is measured with an accuracy of about one part in 10^{11}. One of the most precisely determined quantities in physics, the magnetic dipole moment of the electron can be calculated correctly from quantum theory to within about one part in 10^{10}.

CHAPTER 4
QUANTUM MECHANICS:
INTERPRETATION

Although quantum mechanics has been applied to problems in physics with great success, some of its ideas seem strange. A few of their implications are considered here.

THE ELECTRON: WAVE OR PARTICLE?

Young's aforementioned experiment in which a parallel beam of monochromatic light is passed through a pair of narrow parallel slits has an electron counterpart. In Young's original experiment, the intensity of the light varies with direction after passing through the slits. The intensity oscillates because of interference between the light waves emerging from the two slits, the rate of oscillation depending on the wavelength of the light and the separation of the slits. The oscillation creates a fringe pattern of alternating light and dark bands that is modulated by the diffraction pattern from each slit. If one of the slits is covered, the interference fringes disappear, and only the diffraction pattern is observed.

Young's experiment can be repeated with electrons all with the same momentum. The screen in the optical experiment is replaced by a closely spaced grid of electron detectors. There are many devices for detecting electrons; the most common are scintillators. When an electron passes through a scintillating material, such as sodium iodide, the material produces a light flash which gives a voltage pulse that can be amplified and recorded. The pattern of electrons recorded by each detector is the same

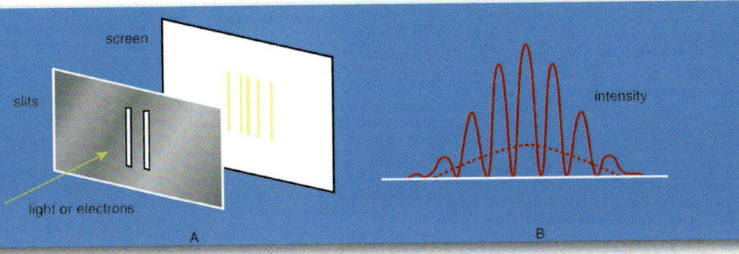

(A) Monochromatic light incident on a pair of slits gives interference fringes (alternate light and dark bands) on a screen, (B) variation in the intensity of the light at the screen when both slits are open. With a single slit, there is no interference pattern; the intensity variation is shown by the broken line. Copyright Encyclopædia Britannica; rendering for this edition by Rosen Educational Services

as that predicted for waves with wavelengths given by the Broglie formula. Thus, the experiment provides conclusive evidence for the wave behaviour of electrons.

If the experiment is repeated with a very weak source of electrons so that only one electron passes through the slits, a single detector registers the arrival of an electron. This is a well-localized event characteristic of a particle. Each time the experiment is repeated, one electron passes through the slits and is detected. A graph plotted with detector position along one axis and the number of electrons along the other looks exactly like the oscillating interference pattern. Thus, the intensity function in the figure is proportional to the probability of the electron moving in a particular direction after it has passed through the slits. Apart from its units, the function is identical to Ψ^2, where Ψ is the solution of the time-independent Schrödinger equation for this particular experiment.

If one of the slits is covered, the fringe pattern disappears and is replaced by the diffraction pattern for a single slit. Thus, both slits are needed to produce the fringe pattern. However, if the electron is a particle, it seems reasonable to suppose that it passed through only one of the slits. The apparatus can be modified to ascertain which slit by placing a thin wire loop around each slit. When an electron passes through a loop, it generates a small electric signal, showing which slit it passed through. However, the interference fringe pattern then disappears, and the single-slit diffraction pattern returns. Since both slits are needed for the interference pattern to appear and since it is impossible to know which slit the electron passed through without destroying that pattern, one is forced to the conclusion that the electron goes through both slits at the same time.

In summary, the experiment shows both the wave and particle properties of the electron. The wave property predicts the probability of direction of travel before the electron is detected; on the other hand, the fact that the electron is detected in a particular place shows that it has particle properties. Therefore, the answer to the question whether the electron is a wave or a particle is that it is neither. It is an object exhibiting either wave or particle properties, depending on the type of measurement that is made on it. In other words, one cannot talk about the intrinsic properties of an electron; instead, one must consider the properties of the electron and measuring apparatus together.

HIDDEN VARIABLES

A fundamental concept in quantum mechanics is that of randomness, or indeterminacy. In general, the theory predicts only the probability of a certain result. Consider the case of radioactivity. Imagine a box of atoms with identical nuclei that can undergo decay with the emission of an alpha particle. In a given time interval, a certain fraction will decay. The theory may tell precisely what that fraction will be, but it cannot predict which particular nuclei will decay. The theory asserts that, at the beginning of the time interval, all the nuclei are in an identical state and that the decay is a completely random process. Even in classical physics, many processes appear random. For example, one says that, when a roulette wheel is spun, the ball will drop at random into one of the numbered compartments in the wheel. Based on this belief, the casino owner and the players give and accept identical odds against each number for each throw. However, the fact is that the winning number could be predicted if one noted the exact location of the wheel when the croupier released the ball, the initial speed of the wheel, and various other physical parameters. It is only ignorance of the initial conditions and the difficulty of doing the calculations that makes the outcome appear to be random. In quantum mechanics, on the other hand, the randomness is asserted to be absolutely fundamental. The theory says that, though one nucleus decayed and the other did not, they were previously in the identical state.

Many eminent physicists, including Einstein, have not accepted this indeterminacy. They have rejected the notion that the nuclei were initially in the identical state. Instead, they postulated that there must be some other property—presently unknown, but existing nonetheless—that is different for the two nuclei. This type of unknown property is termed a hidden variable; if it existed, it would restore determinacy to physics. If the initial values of the hidden variables were known, it would be possible to predict which nuclei would decay. Such a theory would, of course, also have to account for the wealth of experimental data which conventional quantum mechanics explains from a few simple assumptions. Attempts have been made by Broglie, David Bohm, and others to construct theories based on hidden variables, but the theories are very complicated and contrived. For example, the electron would definitely have to go through only one slit in the two-slit experiment. To explain that interference occurs only when the other slit is open, it is necessary to postulate a special force on the electron which exists only when that slit is open. Such artificial additions make hidden variable theories unattractive, and there is little support for them among physicists.

The orthodox view of quantum mechanics—and the one adopted in the present article—is known as the Copenhagen interpretation because its main protagonist, Niels Bohr, worked in that city. The Copenhagen view of understanding the physical world stresses the importance of basing theory on what can be observed and measured experimentally. It therefore rejects the idea of hidden variables as quantities that cannot be measured. The Copenhagen view is that the indeterminacy observed in nature is fundamental and does not reflect an inadequacy in present scientific knowledge. One should therefore accept the indeterminacy without trying to "explain" it and see what consequences come from it.

Attempts have been made to link the existence of free will with the indeterminacy of quantum mechanics, but it is difficult to see how this feature of the theory makes free will more plausible. On the contrary, free will presumably implies rational thought and decision, whereas the essence of the indeterminism in quantum mechanics is that it is due to intrinsic randomness.

PARADOX OF EINSTEIN, PODOLSKY, AND ROSEN

In 1935 Einstein and two other physicists in the United States, Boris Podolsky and Nathan Rosen, analyzed a thought experiment to measure position and momentum in a pair of interacting systems. Employing conventional quantum mechanics, they obtained some startling results, which led them to conclude that the theory does not give a complete description of physical reality. Their results, which are so peculiar as to seem paradoxical, are based on impeccable reasoning, but their conclusion that the theory is incomplete does not necessarily follow. Bohm simplified their experiment while retaining the central point of their reasoning; this discussion follows his account.

The proton, like the electron, has spin ½; thus, no matter what direction is chosen for measuring the component of its spin angular momentum, the values are always $+\hbar/2$ or $-\hbar/2$. (The present discussion relates only to spin angular momentum, and the word spin is omitted from now on.) It is possible to obtain a system consisting of a pair of protons in close proximity and with total angular momentum equal to zero. Thus, if the value of one of the components of angular momentum for one of the protons is $+\hbar/2$ along any selected direction, the value for the component in the same direction for the other particle must be $-\hbar/2$. Suppose the two protons move in opposite directions until they are far apart. The total angular momentum of the system remains zero, and if the component of angular momentum along the same direction for each of the two particles is measured, the result is a pair of equal and opposite values. Therefore, after the quantity is measured for one of the protons, it can be predicted for the other proton; the second measurement is unnecessary. As previously noted, measuring a quantity changes the state of the system. Thus, if measuring S_x (the x-component of angular momentum) for proton 1 produces the value $+\hbar/2$, the state of proton 1 after measurement corresponds to $S_x = +\hbar/2$, and the state of proton 2 corresponds to $S_x = -\hbar/2$. Any direction, however, can be chosen for measuring the component of angular momentum. Whichever direction is selected, the state of proton 1 after measurement corresponds to a definite component of angular momentum about that direction. Furthermore, since proton 2 must have the opposite value for the same component, it follows that the measurement on proton 1 results in a definite state for proton 2 relative to the chosen direction, notwithstanding the fact that the two particles may be millions of kilometres apart and are not interacting with each other at the time. Einstein and his two collaborators thought that this conclusion was so obviously false that the quantum mechanical theory on which it was based must be incomplete. They concluded that the correct theory would contain some hidden variable feature that would restore the determinism of classical physics.

A comparison of how quantum theory and classical theory describe angular momentum for particle pairs illustrates the essential difference between the two outlooks. In both theories, if a system of two particles has a total angular momentum of zero, then the angular momenta of the two particles are equal and opposite. If the components of angular momentum are measured along the same direction, the two values are numerically equal, one positive and the other negative. Thus, if one component is measured, the other can be predicted. The crucial difference between the two theories is that, in classical physics, the system under investigation is assumed to have possessed the quantity being measured beforehand. The measurement does not disturb the system; it merely reveals the preexisting state. It may be noted that, if a particle were actually to possess components of angular momentum prior to measurement, such quantities would constitute hidden variables.

Does nature behave as quantum mechanics predicts? The answer comes from measuring the components of angular momenta for the two protons along different directions with an angle θ between them. A measurement on one proton can give only the result $+\hbar/2$ or $-\hbar/2$. The experiment consists of measuring correlations between

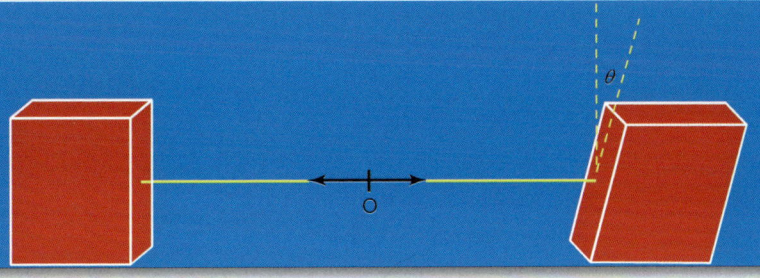

Experiment to determine the correlation in measured angular momentum values for a pair of protons with zero total angular momentum. The two protons are initially at the point o and move in opposite directions toward the two magnets. Copyright Encyclopædia Britannica; rendering for this edition by Rosen Educational Services

the plus and minus values for pairs of protons with a fixed value of θ, and then repeating the measurements for different values of θ. The interpretation of the results rests on an important theorem by the British physicist John Stewart Bell. Bell began by assuming the existence of some form of hidden variable with a value that would determine whether the measured angular momentum gives a plus or minus result. He further assumed locality—namely, that measurement on one proton (i.e., the choice of the measurement direction) cannot affect the result of the measurement on the other proton. Both these assumptions agree with classical, commonsense ideas. He then showed quite generally that these two assumptions lead to a certain relationship, now known as Bell's inequality, for the correlation values mentioned above. Experiments have been conducted at several laboratories with photons instead of protons (the analysis is similar), and the results show fairly conclusively that Bell's inequality is violated. That is to say, the observed results agree with those of quantum mechanics and cannot be accounted for by a hidden variable (or deterministic) theory based on the concept of locality. One is forced to conclude that the two protons are a correlated pair and that a measurement on one affects the state of both, no matter how far apart they are. This may strike one as highly peculiar, but such is the way nature appears to be.

It may be noted that the effect on the state of proton 2 following a measurement on proton 1 is believed to be instantaneous; the effect happens before a light signal initiated by the measuring event at proton 1 reaches proton 2. Alain Aspect and his coworkers in Paris demonstrated this result in 1982 with an ingenious experiment in which the correlation between the two angular momenta was measured, within a very short time interval, by a high-frequency switching device. The interval was less than the time taken for a light signal to travel from one particle to the other at the two measurement positions. Einstein's special theory of relativity states that no message can travel with a speed greater than that of light. Thus, there is no way that the information concerning the direction of the measurement on the first proton could reach the second proton before the measurement was made on it.

MEASUREMENT IN QUANTUM MECHANICS

The way quantum mechanics treats the process of measurement has caused considerable debate. Schrödinger's time-dependent wave equation is an exact recipe for determining the way the wave function varies with time for a given physical system in a given physical environment. According to the Schrödinger equation, the wave function varies in a strictly determinate way. On the other hand, in the axiomatic approach to quantum mechanics described above, a measurement changes the wave function abruptly and discontinuously. Before the measurement is made, the wave function Ψ is a mixture of the states ψs. The measurement changes Ψ from a mixture of

ψs to a single ψ. This change, brought about by the process of measurement, is termed the collapse or reduction of the wave function. The collapse is a discontinuous change in Ψ; it is also unpredictable, because, starting with the same Ψ, the end result can be any one of the individual ψs.

The Schrödinger equation, which gives a smooth and predictable variation of Ψ, applies between the measurements. The measurement process itself, however, cannot be described by the Schrödinger equation; it is somehow a thing apart. This appears unsatisfactory, inasmuch as a measurement is a physical process and ought to be the subject of the Schrödinger equation just like any other physical process.

The difficulty is related to the fact that quantum mechanics applies to microscopic systems containing one (or a few) electrons, protons, or photons. Measurements, however, are made with large-scale objects (e.g., detectors, amplifiers, and meters) in the macroscopic world, which obeys the laws of classical physics. Thus, another way of formulating the question of what happens in a measurement is to ask how the microscopic quantum world relates and interacts with the macroscopic classical world. More narrowly, it can be asked how and at what point in the measurement process does the wave function collapse? So far, there are no satisfactory answers to these questions, although there are several schools of thought.

One approach stresses the role of a conscious observer in the measurement process and suggests that the wave function collapses when the observer reads the measuring instrument. Bringing the conscious mind into the measurement problem seems to raise more questions than it answers, however.

As discussed above, the Copenhagen interpretation of the measurement process is essentially pragmatic. It distinguishes between microscopic quantum systems and macroscopic measuring instruments. The initial object or event—e.g., the passage of an electron, photon, or atom—triggers the classical measuring device into giving a reading; somewhere along the chain of events, the result of the measurement becomes fixed (i.e., the wave function collapses). This does not answer the basic question but says, in effect, not to worry about it. This is probably the view of most practicing physicists.

A third school of thought notes that an essential feature of the measuring process is irreversibility. This contrasts with the behaviour of the wave function when it varies according to the Schrödinger equation; in principle, any such variation in the wave function can be reversed by an appropriate experimental arrangement. However, once a classical measuring instrument has given a reading, the process is not reversible. It is possible that the key to the nature of the measurement process lies somewhere here. The Schrödinger equation is known to apply only to relatively simple systems. It is an enormous extrapolation to assume that the same equation applies to the large and complex system of a classical measuring device. It may be that the appropriate equation for such a system has features that produce irreversible effects (e.g., wave-function collapse) which differ in kind from those for a simple system.

One may also mention the so-called many-worlds interpretation, proposed by Hugh Everett III in 1957, which suggests that, when a measurement is made for a system in which the wave function is a mixture of states, the universe branches into a number of noninteracting universes. Each of the possible outcomes of the measurement occurs, but in a different universe. Thus, if $S_x = ½$ is the result of a Stern-Gerlach measurement on a silver atom, there is another universe identical to ours in every way (including clones of people), except that the result of the measurement is $S_x = -½$. Although this fanciful model solves some measurement problems, it has few adherents among physicists.

Because the various ways of looking at the measurement process lead to the same experimental consequences, trying to distinguish between them on scientific grounds may be fruitless. One or another may be preferred on the grounds of plausibility, elegance, or economy of hypotheses, but these are matters of individual taste. Whether one day a satisfactory quantum theory of measurement will emerge, distinguished from the others by its verifiable predictions, remains an open question.

APPLICATIONS OF QUANTUM MECHANICS

As has been noted, quantum mechanics has been enormously successful in explaining microscopic phenomena in all branches of physics. The three phenomena described in this section are examples that demonstrate the quintessence of the theory.

DECAY OF A MESON

The $K°$ meson, discovered in 1953, is produced in high-energy collisions between nuclei and other particles. It has zero electric charge, and its mass is about one-half the mass of the proton. It is unstable and, once formed, rapidly decays into either 2 or 3 pi-mesons. The average lifetime of the $K°$ is about 10^{-10} second.

In spite of the fact that the $K°$ meson is uncharged, quantum theory predicts the existence of an antiparticle with the same mass, decay products, and average lifetime; the antiparticle is denoted by $\overline{K°}$. During the early 1950s, several physicists questioned the justification for postulating the existence of two particles with such similar properties. In 1955, however, Murray Gell-Mann and Abraham Pais made an interesting prediction about the decay of the $K°$ meson. Their reasoning provides an excellent illustration of the quantum mechanical axiom that the wave function Ψ can be a superposition of states; in this case, there are two states, the $K°$ and $\overline{K°}$ mesons themselves.

A $K°$ meson may be represented formally by writing the wave function as $\Psi = K°$; similarly $\Psi = \overline{K°}$ represents a $\overline{K°}$ meson. From the two states, $K°$ and $\overline{K°}$, the following two new states are constructed:

$$K_1 = \frac{(K° + \overline{K°})}{\sqrt{2}} \qquad (14)$$

$$K_2 = \frac{(K° - \overline{K°})}{\sqrt{2}}. \qquad (15)$$

From these two equations it follows that

$$K° = \frac{(K_1 + K_2)}{\sqrt{2}}, \qquad (\qquad (16)$$

$$\overline{K°} = \frac{(K_1 - K_2)}{\sqrt{2}}. \qquad (17)$$

The reason for defining the two states K_1 and K_2 is that, according to quantum theory, when the $K°$ decays, it does not do so as an isolated particle; instead, it combines with its antiparticle to form the states K_1 and K_2. The state K_1 decays into two pi-mesons with a very short lifetime (about 10^{-10} second), while K_2 decays into three pi-mesons with a longer lifetime (about 10^{-7} second).

The physical consequences of these results may be demonstrated in the following experiment. $K°$ particles are produced in a nuclear reaction at the point A. They move to the right and start to decay. At point A, the wave function is $\Psi = K°$, which, from equation (16), can be expressed as the sum of K_1 and K_2. As the particles move to the right, the K_1 state begins to decay rapidly. If the particles reach point B in about 10^{-8} second, nearly all the K_1 component has decayed, although hardly any of the K_2 component has done so. Thus, at point B, the beam has changed from one of pure $K°$ to one of almost pure K_2, which equation (15) shows is an equal mixture of $K°$ and $\overline{K°}$. In other words, $\overline{K°}$ particles appear in the beam simply because K_1 and K_2 decay at different rates. At point B, the beam enters a block of absorbing material. Both the $K°$ and $\overline{K°}$ are absorbed by the nuclei in the block, but the $\overline{K°}$ are absorbed more strongly. As a result, even though the beam is an equal mixture of $K°$ and $\overline{K°}$ when it enters the absorber, it is almost pure $K°$ when it exits at point C. The beam thus begins and ends as $K°$.

Gell-Mann and Pais predicted all this, and experiments subsequently verified it. The experimental observations are that the decay products are primarily two pi-mesons with a short decay time near A, three pi-mesons with longer decay time near B, and two pi-mesons again near C. (This account exaggerates the changes in the K_1 and K_2 components between A and B and in the $K°$ and $\overline{K°}$ components between B and C; the argument, however, is unchanged.) The phenomenon of generating the $\overline{K°}$ and regenerating the K_1 decay is purely quantum. It rests on

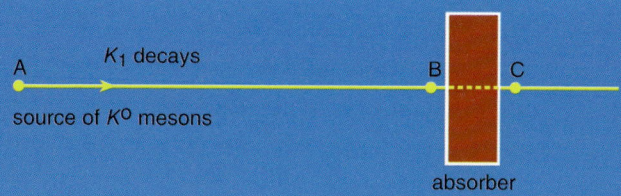

Decay of the $K°$ meson. Copyright Encyclopædia Britannica; rendering for this edition by Rosen Educational Services

the quantum axiom of the superposition of states and has no classical counterpart.

Cesium Clock

The cesium clock is the most accurate type of clock yet developed. This device makes use of transitions between the spin states of the cesium nucleus and produces a frequency which is so regular that it has been adopted for establishing the time standard.

Like electrons, many atomic nuclei have spin. The spin of these nuclei produces a set of small effects in the spectra, known as hyperfine structure. (The effects are small because, though the angular momentum of a spinning nucleus is of the same magnitude as that of an electron, its magnetic moment, which governs the energies of the atomic levels, is relatively small.) The nucleus of the cesium atom has spin quantum number $7/2$. The total angular momentum of the lowest energy states of the cesium atom is obtained by combining the spin angular momentum of the nucleus with that of the single valence electron in the atom. (Only the valence electron contributes to the angular momentum because the angular momenta of all the other electrons total zero. Another simplifying feature is that the ground states have zero orbital momenta, so only spin angular momenta need to be considered.) When nuclear spin is taken into account, the total angular momentum of the atom is characterized by a quantum number, conventionally denoted by F, which for cesium is 4 or 3. These values come from the spin value $7/2$ for the nucleus and $½$ for the electron. If the nucleus and the electron are visualized as tiny spinning tops, the value $F = 4$ ($7/2 + ½$) corresponds to the tops spinning in the same sense, and $F = 3$ ($7/2 - ½$) corresponds to spins in opposite senses. The energy difference QE of the states with the two F values is a precise quantity. If electromagnetic radiation of frequency v_0, where

$$hv_0 = \Delta E, \qquad (18)$$

is applied to a system of cesium atoms, transitions will occur between the two states. An apparatus that can detect the occurrence of transitions thus provides an extremely precise frequency standard. This is the principle of the cesium clock.

A beam of cesium atoms emerges from an oven at a temperature of about 100 °C (212 °F). The atoms pass through an inhomogeneous magnet A, which deflects the atoms in state $F = 4$ downward and those in state $F = 3$ by an equal amount upward. The atoms pass through slit S and continue into a second inhomogeneous magnet B. Magnet B is arranged so that it deflects atoms with an unchanged state in the same direction that magnet A deflected them. The atoms follow the paths indicated by the broken lines in the figure and are lost to the beam. However, if an alternating electromagnetic field of frequency v_0 is applied to the beam as it traverses the centre region C, transitions between states will occur. Some atoms in state $F = 4$ will change to $F = 3$, and vice versa. For such atoms, the deflections in magnet B are reversed. The atoms follow the whole lines in the diagram and strike a tungsten wire,

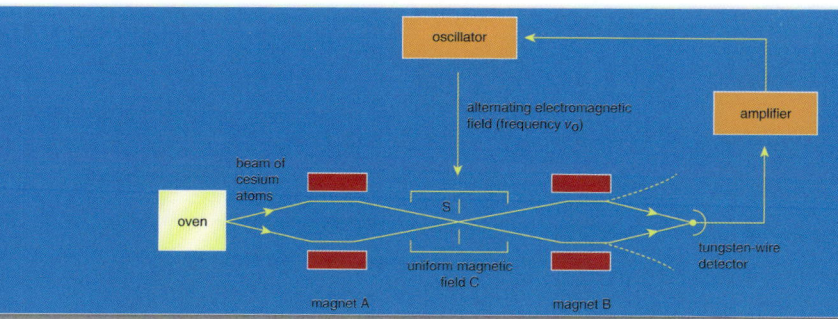

Cesium clock. Copyright Encyclopædia Britannica; rendering for this edition by Rosen Educational Services

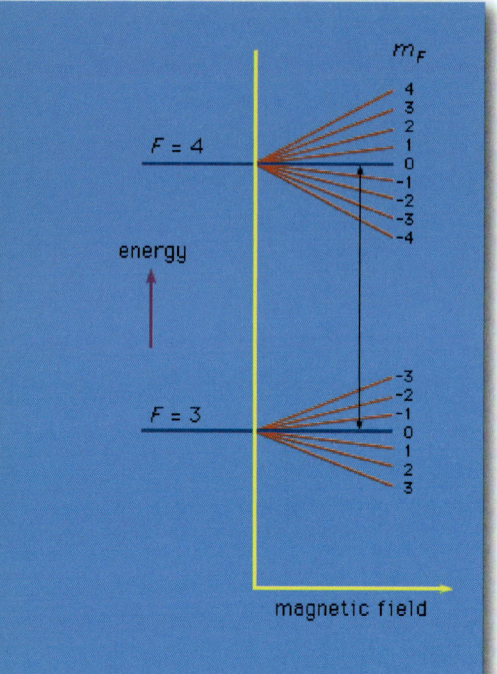

Variation of energy with magnetic-field strength for the F = 4 and F = 3 states in cesium-133. Copyright Encyclopædia Britannica; rendering for this edition by Rosen Educational Services

which gives electric signals in proportion to the number of cesium atoms striking the wire. As the frequency v of the alternating field is varied, the signal has a sharp maximum for $v = v_o$. The length of the apparatus from the oven to the tungsten detector is about one metre.

Each atomic state is characterized not only by the quantum number F but also by a second quantum number m_F. For $F = 4$, m_F can take integral values from 4 to -4. In the absence of a magnetic field, these states have the same energy. A magnetic field, however, causes a small change in energy proportional to the magnitude of the field and to the m_F value. Similarly, a magnetic field changes the energy for the $F = 3$ states according to the m_F value which, in this case, may vary from 3 to -3. In the cesium clock, a weak constant magnetic field is superposed on the alternating electromagnetic field in region C. The theory shows that the alternating field can bring about a transition only between pairs of states with m_F values that are the same or that differ by unity. However, the only transitions occurring at the frequency v_o are those between the two states with $m_F = 0$. The apparatus is so sensitive that it can discriminate easily between such transitions and all the others.

If the frequency of the oscillator drifts slightly so that it does not quite equal v_o, the detector output drops. The change in signal strength produces a signal to the oscillator to bring the frequency back to the correct value. This feedback system keeps the oscillator frequency automatically locked to v_o.

The cesium clock is exceedingly stable. The frequency of the oscillator remains constant to about one part in 10^{13}. For this reason, the device has been used to redefine the second. This base unit of time in the SI system is defined as equal to 9,192,631,770 cycles of the radiation corresponding to the transition between the levels $F = 4$, $m_F = 0$ and $F = 3, m_F = 0$ of the ground state of the cesium-133 atom. Prior to 1964, the second was defined in terms of the motion of Earth. The latter, however, is not nearly as stable as the cesium clock. Specifically, the fractional variation of Earth's rotation period is a few hundred times larger than that of the frequency of the cesium clock.

A Quantum Voltage Standard

Quantum theory has been used to establish a voltage standard, and this standard has proven to be extraordinarily accurate and consistent from laboratory to laboratory.

If two layers of superconducting material are separated by a thin insulating barrier, a supercurrent (i.e., a current of paired electrons) can pass from one superconductor to the other. This is another example of the tunneling process described earlier. Several effects based on this phenomenon were predicted in 1962 by the British physicist Brian D. Josephson. Demonstrated experimentally soon afterwards, they are now referred to as the Josephson effects.

If a DC (direct-current) voltage V is applied across the two superconductors, the energy of an electron pair changes by an amount of $2eV$ as it crosses the junction. As a result, the supercurrent oscillates with frequency ν given by the Planck relationship ($E = h\nu$). Thus,

$$2eV = h\nu \qquad (19)$$

This oscillatory behaviour of the supercurrent is known as the AC (alternating-current) Josephson effect. Measurement of V and ν permits a direct verification of the Planck relationship. Although the oscillating supercurrent has been detected directly, it is extremely weak. A more sensitive method of investigating equation (19) is to study effects resulting from the interaction of microwave radiation with the supercurrent.

Several carefully conducted experiments have verified equation (19) to such a high degree of precision that it has been used to determine the value of $2e/h$. This value can in fact be determined more precisely by the AC Josephson effect than by any other method. The result is so reliable that laboratories now employ the AC Josephson effect to set a voltage standard. The numerical relationship between V and ν is

$$\frac{2e}{h} = \frac{\nu}{V} = 483{,}597.7 \times 10^9 \text{ hertz per volt} \qquad (20)$$

In this way, measuring a frequency, which can be done with great precision, gives the value of the voltage. Before the Josephson method was used, the voltage standard in metrological laboratories devoted to the maintenance of physical units was based on high-stability Weston cadmium cells. These cells, however, tend to drift and so caused inconsistencies between standards in different laboratories. The Josephson method has provided a standard giving agreement to within a few parts in 10^8 for measurements made at different times and in different laboratories.

The experiments described in the preceding two sections are only two examples of high-precision measurements in physics. The values of the fundamental constants, such as c, h, e, and m_e, are determined from a wide variety of experiments based on quantum phenomena. The results are so consistent that the values of the constants are thought to be known in most cases to better than one part in 10^6. Physicists may not know what they are doing when they make a measurement, but they do it extremely well.

Bose-Einstein Condensate

The Bose-Einstein condensate (BEC) is a state of matter in which separate atoms or subatomic particles, cooled to near absolute zero (0 K, -273.15 °C, or -459.67 °F; K = kelvin), coalesce into a single quantum mechanical entity—that is, one that can be described by a wave function—on a near-macroscopic scale. This form of matter was predicted in 1924 by Albert Einstein on the basis of the quantum formulations of the Indian physicist Satyendra Nath Bose.

Although it had been predicted for decades, the first atomic BEC was made only in 1995, when Eric Cornell and Carl Wieman of JILA, a research institution jointly operated by the National Institute of Standards and Technology (NIST) and the University of Colorado at Boulder, cooled a gas of rubidium atoms to 1.7×10^{-7} K above absolute zero. Along with Wolfgang Ketterle of the Massachusetts Institute of Technology (MIT), who created a BEC with sodium atoms, these researchers received the 2001 Nobel Prize for Physics. Research on BECs has expanded the understanding of quantum physics and has led to the discovery of new physical effects.

BEC theory traces back to 1924, when Bose considered how groups of photons behave. Photons belong to one of the two great classes of elementary or submicroscopic particles defined by whether their quantum spin is a nonnegative integer (0, 1, 2, ...) or an odd half integer (1/2, 3/2, ...). The former type, called bosons, includes photons, whose spin is 1. The latter type, called fermions, includes electrons, whose spin is 1/2.

As Bose noted, the two classes behave differently. According to the Pauli exclusion principle, fermions tend to avoid each other, for which reason each electron in a group occupies a separate quantum state (indicated by different quantum numbers, such as the electron's energy). In contrast, an unlimited number of bosons can have the same energy state and share a single quantum state.

Einstein soon extended Bose's work to show that at extremely low temperatures "bosonic atoms" with even spins would coalesce into a shared quantum state at the lowest available energy. (Thus, bosons are said to follow Bose-Einstein statistics, and fermions follow Fermi-Dirac statistics.) The requisite methods to produce temperatures low enough to test Einstein's prediction did not become attainable, however, until the 1990s. One of the breakthroughs depended on the novel technique of laser cooling and trapping, in which the radiation pressure of a laser beam cools and localizes atoms by slowing them down. (For this work, French physicist Claude Cohen-Tannoudji and American physicists Steven Chu and William D. Phillips shared the 1997 Nobel Prize for Physics.) The second breakthrough depended on improvements in magnetic confinement in order to hold the atoms in place without a material container. Using these techniques, Cornell and Wieman succeeded in merging about 2,000 individual atoms into a "superatom," a condensate large enough to observe with a microscope, that displayed distinct quantum properties. As Wieman described the achievement, "We brought it to an almost human scale. We can poke it and prod it and look at this stuff in a way no one has been able to before."

BECs are related to two remarkable low-temperature phenomena: superfluidity, in which each of the helium isotopes ^{3}He and ^{4}He forms a liquid that flows with zero friction; and superconductivity, in which electrons move through a material with zero electrical resistance. ^{4}He atoms are bosons, and although ^{3}He atoms and electrons are fermions, they can also undergo Bose condensation if they pair up with opposite spins to form bosonlike states with zero net spin. In 2003 Deborah Jin and her colleagues at JILA used paired fermions to create the first atomic fermionic condensate.

BEC research has yielded new atomic and optical physics, such as the atom laser Ketterle demonstrated in 1996. A conventional light laser emits a beam of coherent photons; they are all exactly in phase and can be focused to an extremely small, bright spot. Similarly, an atom laser produces a coherent beam of atoms that can be focused at high intensity. Potential applications include more-accurate atomic clocks and enhanced techniques to make electronic chips, or integrated circuits.

The most intriguing property of BECs is that they can slow down light. In 1998 Lene Hau of Harvard University and her colleagues slowed light traveling through a BEC from its speed in vacuum of 3×10^8 metres per second to a mere 17 metres per second, or about 38 miles per hour. Since then, Hau and others have completely halted and stored a light pulse within a BEC, later releasing the light unchanged or sending it to a second BEC. These manipulations hold promise for new types of light-based telecommunications, optical storage of data, and quantum computing, though the low-temperature requirements of BECs offer practical difficulties.

CHAPTER 5
BIOGRAPHIES

The 20th century was a time of great ferment in physics with the arrival of both relativity and quantum mechanics. This section presents the biographies of many of the scientists who founded these theories.

CARL DAVID ANDERSON

(b. Sept. 3, 1905, New York, N.Y., U.S.—d. Jan. 11, 1991, San Marino, Calif.)

American physicist Carl David Anderson, with Victor Francis Hess of Austria, won the Nobel Prize for Physics in 1936 for his discovery of the positron, or positive electron, the first known particle of antimatter.

Anderson received his Ph.D. in 1930 from the California Institute of Technology, Pasadena, where he worked with physicist Robert Andrews Millikan. Having studied X-ray photoelectrons (electrons ejected from atoms by interaction with high-energy photons) since 1927, he began research in 1930 on gamma rays and cosmic rays. While studying cloud-chamber photographs of cosmic rays, Anderson found a number of tracks whose orientation suggested that they were caused by positively charged particles—but particles too small to be protons. In 1932 he announced that they were caused by positrons, positively charged particles with the same mass as electrons. The claim was controversial until verified the next year by British physicist Patrick M.S. Blackett and Italian Giuseppe Occhialini.

In 1936 Anderson discovered the mu-meson, or muon, a subatomic particle 207 times heavier than the electron. At first he thought he had found the meson, postulated by the Japanese physicist Jukawa Hideki, that binds protons and neutrons together in the nucleus of the atom, but the muon was found to interact weakly with these particles. (The particle predicted by Yukawa was discovered in 1947 by the British physicist Cecil Powell and is known as a pi-meson, or pion.)

Anderson spent his entire career at Caltech, joining the faculty in 1933 and serving as professor until 1976. During World War II he conducted research on rockets.

HANS BETHE

(b. July 2, 1906, Strassburg, Ger. [now Strasbourg, France]—d. March 6, 2005, Ithaca, N.Y., U.S.)

German-born American theoretical physicist Hans Albrecht Bethe helped shape quantum physics and increased the understanding of the atomic processes responsible for the properties of matter and of the forces governing the structures of atomic nuclei. He received the Nobel Prize for Physics in 1967 for his work on the production of energy in stars. Moreover, he was a leader in emphasizing the social responsibility of science.

Bethe started reading at age four and began writing at about the same age. His numerical and mathematical abilities also manifested themselves early. His mathematics teacher at the local gymnasium recognized his talents and encouraged him to continue studies in mathematics and the physical sciences. Bethe graduated from the gymnasium in the spring of 1924. After completing two years of studies at the University of Frankfurt, he was advised by one of his teachers to go

Hans Bethe. *SPL/Photo Researchers, Inc.*

to the University of Munich and study with Arnold Sommerfeld.

It was in Munich that Bethe discovered his exceptional proficiency in physics. Sommerfeld indicated to him that he was among the very best students who had studied with him, and these included Wolfgang Pauli and Werner Heisenberg. Bethe obtained a doctorate in 1928 with a thesis on electron diffraction in crystals. During 1930, as a Rockefeller Foundation fellow, Bethe spent a semester at the University of Cambridge under the aegis of Ralph Fowler and a semester at the University of Rome working with Enrico Fermi.

Bethe's craftsmanship was an amalgam of what he had learned from Sommerfeld and from Fermi, combining the best of both: the thoroughness and rigor of Sommerfeld and the clarity and simplicity of Fermi. This craftsmanship was displayed in full force in the many reviews that Bethe wrote. His two book-length reviews in the 1933 *Handbuch der Physik*—the first with Sommerfeld on solid-state physics and the second on the quantum theory of one- and two-electron systems—exhibited his remarkable powers of synthesis. Along with a review on nuclear physics in *Reviews of Modern Physics* (1936–37), these works were instant classics. All of Bethe's reviews were syntheses of the fields under review, giving them coherence and unity while charting the paths to be taken in addressing new problems. They usually contained much new material that Bethe had worked out in their preparation.

In the fall of 1932, Bethe obtained an appointment at the University of Tübingen as an acting assistant professor of theoretical physics. In April 1933, after Adolf Hitler's accession to power, he was dismissed because his maternal grandparents were Jews. Sommerfeld was able to help him by awarding him a fellowship for the summer of 1933, and he got William Lawrence Bragg to invite him to the University of Manchester, Eng., for the following academic year. Bethe then went to the University of Bristol for the 1934 fall semester before accepting a position at Cornell University, Ithaca, N.Y. He arrived at Cornell in February 1935, and he stayed there for the rest of his life.

Bethe came to the United States at a time when the American physics community was undergoing enormous growth. The Washington Conferences on Theoretical Physics were paradigmatic of the meetings organized to assimilate the insights quantum mechanics was giving to many fields, especially atomic and molecular physics and the emerging field of nuclear physics. Bethe attended the 1935 and 1937 Washington Conferences, but he agreed to participate in the 1938 conference on stellar energy generation only after repeated urgings by Edward Teller. As a result of what he learned at the latter conference, Bethe was able to give definitive answers to the problem of energy generation in stars. By stipulating and analyzing the nuclear reactions responsible for the phenomenon, he explained how stars could continue to burn for billions of years. His 1939 *Physical Review* paper on energy generation in stars created the field of nuclear astrophysics and led to his being awarded the Nobel Prize.

During World War II Bethe first worked on problems in radar, spending a year at the Radiation Laboratory at the Massachusetts Institute of Technology. In 1943 he joined the Los Alamos Laboratory (now the Los Alamos National Laboratory) in New Mexico as the head of its theoretical division. He and the division were part of the Manhattan Project, and they made crucial contributions to the feasibility and design of the uranium and the plutonium atomic bombs. The years at Los Alamos changed his life.

In the aftermath of the development of these fission weapons, Bethe became deeply involved with investigating the feasibility of developing fusion bombs, hoping to prove that no terrestrial mechanism could accomplish the task. He believed their development to be immoral. When the Teller-Ulam mechanism for igniting a fusion reaction was advanced in 1951 and the possibility of a hydrogen bomb, or H-bomb, became a reality, Bethe helped to design it. He believed that the Soviets would likewise be able to build one and that only a balance of terror would prevent their use.

As a result of these activities, Bethe became deeply occupied with what he called "political physics," the attempt to educate the public and politicians about the consequences of the existence of nuclear weapons. He became a relentless champion of nuclear arms control, writing many essays (collected in *The Road from Los Alamos* [1991]). He also became deeply committed to making peaceful applications of nuclear power economical and safe. Throughout his life, Bethe was a staunch advocate of nuclear power, defending it as an answer to the inevitable exhaustion of fossil fuels.

Bethe served on numerous advisory committees to the United States government, including the President's Science Advisory Committee (PSAC). As a member of PSAC, he helped persuade President Dwight D. Eisenhower to commit the United States to ban atmospheric nuclear tests. (The Nuclear Test Ban Treaty, which banned atmospheric nuclear testing, was finally ratified in 1963.) In 1972 Bethe's cogent and persuasive arguments helped prevent the deployment of antiballistic missile systems. He was influential in opposing President Ronald Reagan's Strategic Defense Initiative, arguing that a space-based laser defense system could be easily countered and that it would lead to further arms escalation. By virtue of these activities, and his general comportment, Bethe became the science community's conscience. It was indicative of Bethe's constant grappling with moral issues that in 1995 he urged fellow scientists to collectively take a "Hippocratic oath" not to work on designing new nuclear weapons.

Throughout the political activism that marked his later life, Bethe never abandoned his scientific researches. Until well into his 90s, he made important contributions at the frontiers of physics and astrophysics. He helped elucidate the properties of neutrinos and explained the observed rate of neutrino emission by the Sun. With the American physicist Gerald Brown, he worked to understand why massive old stars can suddenly become supernovas.

DAVID BOHM

(b. Dec. 20, 1917, Wilkes-Barre, Pa., U.S.—d. Oct. 27, 1992, London, Eng.)

David Bohm was an American-born British theoretical physicist who developed a causal, nonlocal interpretation of quantum mechanics.

Born to an immigrant Jewish family, Bohm defied his father's wishes that he pursue some practical occupation, such as joining the family's furniture business, in order to study science. After receiving a bachelor's degree (1939) from Pennsylvania State College, Bohm continued graduate research at the California Institute of Technology and then the University of California at Berkeley (Ph.D., 1943), where he worked with physicist J. Robert Oppenheimer. In 1947 Bohm became an assistant professor at Princeton University.

In 1943 Bohm was denied security clearance to work at Los Alamos, N.M., on the atomic bomb. His research in Berkeley still proved marginally useful to the Manhattan Project and directed his attention to plasma physics. In postwar papers, Bohm laid the foundations of modern plasma theory. Bohm's lectures at Princeton developed into an influential textbook, *Quantum Theory* (1951), that contained a clear presentation of Danish physicist Niels Bohr's Copenhagen interpretation of quantum mechanics. While working on that book, Bohm came to believe that a causal (non-Copenhagen) interpretation was also possible, contrary to the view then almost universally held among physicists. Encouraged in this pursuit by conversations with Albert Einstein, he developed an interpretation on the assumption that there existed unobserved hidden variables.

By the time his theory was published in 1952, political problems had forced Bohm to emigrate. He had been involved in left-wing politics in Berkeley during World War II, including membership in various organizations that Federal Bureau of Investigations director J. Edgar Hoover labeled communist fronts, which in the postwar climate of McCarthyism made him be seen as a security threat. Bohm refused to testify about his or others' political beliefs to the House Committee on Un-American Activities in 1949, which resulted in his being charged with contempt of the U.S. Congress. Although Bohm was eventually acquitted of the charge, he was suspended from teaching duties and in 1951 lost his job at Princeton. With Einstein's help, he found a position at the University of São Paulo in Brazil and in 1955 at the Technion in Haifa, Israel. After 1957 he worked in England, first at the University of Bristol and then, from 1961 until retirement in 1987, as a professor of theoretical physics at Birkbeck College, University of London.

Initially ignored, the idea of hidden variables inspired interest after the publication of Bohm's *Causality and Chance in Modern Physics* (1957), the prediction of the Aharonov-Bohm effect (1959), and especially after it led American physicist John Bell to discover the Bell inequality theorem (1964). Efforts to interpret quantum theory changed as a result of Bohm's work, with discussion shifting to the issues of nonlocality, nonseparability, and entanglement.

Bohm's later publications became increasingly philosophical; the influence of Marxism on him gave way first to Hegelianism and then theosophy through the teachings of the Indian mystic Jiddu Krishnamurti, with whom he wrote *The Ending of Time* (1985). Bohm's most famous later book, *Wholeness and the Implicate Order* (1980), also dealt with the broader issues of the human condition and consciousness.

NIELS BOHR

(b. Oct. 7, 1885, Copenhagen, Den.—d. Nov. 18, 1962, Copenhagen)

Danish physicist Niels Henrik David Bohr is generally regarded as one of the foremost physicists of the 20th century. He was the first to apply the quantum concept, which restricts the energy of a system to certain discrete values, to the problem of atomic and molecular structure. For this work he received the Nobel Prize for Physics in 1922. His manifold roles in the origins and development of quantum physics may be his most important contribution, but through his long career his involvements were substantially broader, both inside and outside the world of physics.

Bohr was the second of three children born into an upper middle-class Copenhagen family. His mother, Ellen (née Adler), was the daughter of a prominent Jewish banker. His father, Christian, became a professor of physiology at the University of Copenhagen and was nominated twice for the Nobel Prize.

Enrolling at the University of Copenhagen in 1903, Bohr was never in doubt that he would study physics. Research and teaching in this field took place in cramped quarters at the Polytechnic Institute, leased to the University for the purpose. Bohr obtained his doctorate in 1911 with a dissertation on the electron theory of metals.

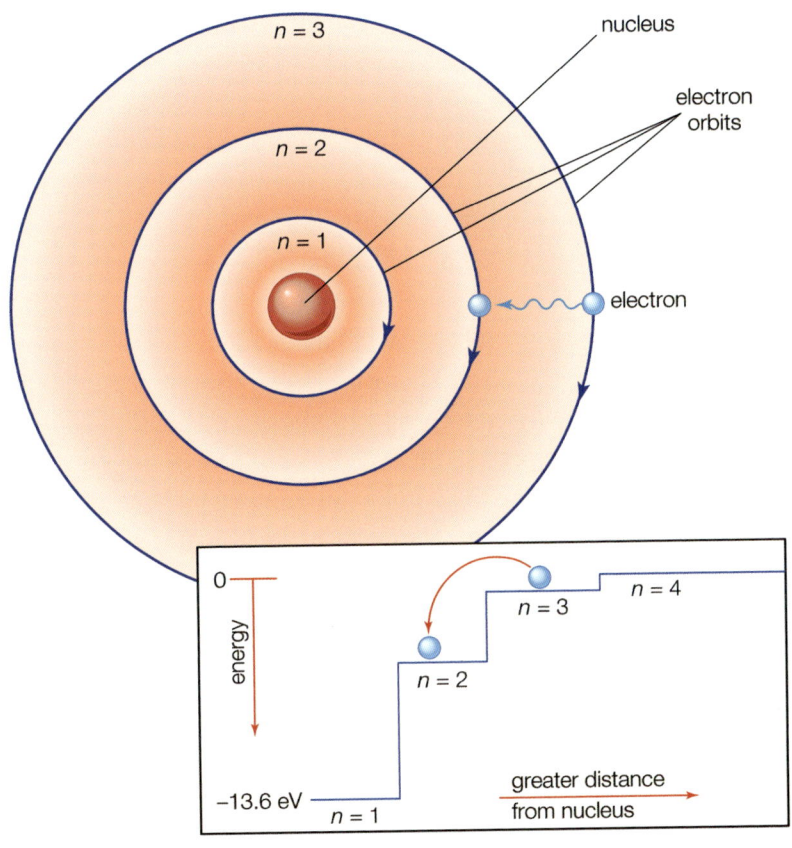

In the Bohr model of the atom, electrons travel in defined circular orbits around the nucleus. The orbits are labeled by an integer, the quantum number n. Electrons can jump from one orbit to another by emitting or absorbing energy. The inset shows an electron jumping from orbit n=3 to orbit n=2, emitting a photon of red light with an energy of 1.89 eV. Encyclopædia Britannica, Inc.

On Aug. 1, 1912, Bohr married Margrethe Nørlund, and the marriage proved a particularly happy one. Throughout his life, Margrethe was his most trusted adviser. They had six sons, the fourth of whom, Aage N. Bohr, shared a third of the 1975 Nobel Prize for Physics in recognition of the collective model of the atomic nucleus proposed in the early 1950s.

Bohr's first contribution to the emerging new idea of quantum physics started in 1912 during what today would be called postdoctoral research in England with Ernest Rutherford at the University of Manchester. Only the year before, Rutherford and his collaborators had established experimentally that the atom consists of a heavy positively charged nucleus with substantially lighter negatively charged electrons circling around it at considerable distance. According to classical physics, such a system would be unstable, and Bohr felt compelled to postulate, in a substantive trilogy of articles published in *The Philosophical Magazine* in 1913, that electrons could only occupy particular orbits determined by the quantum of action and that electromagnetic radiation from an atom occurred only when an electron jumped to a lower-energy orbit. Although radical and unacceptable to most physicists at the time, the Bohr atomic model was able to account for an ever-increasing number of experimental data, famously starting with the spectral line series emitted by hydrogen.

In the spring of 1916, Bohr was offered a new professorship at the University of Copenhagen; dedicated to theoretical physics, it was the second professorship in physics there. As physics was still pursued in the cramped quarters of the Polytechnic Institute, it is not surprising that in the spring of 1917 Bohr wrote a long letter to his faculty asking for the establishment of an Institute for Theoretical Physics. In the inauguration speech for his new institute on March 3, 1921, he stressed, first, that experiments and experimenters were indispensable at an institute for theoretical physics in order to test the statements of the theorists. Second, he expressed his ambition to make the new institute a place where the younger generation of physicists could propose fresh ideas. Starting out with a small staff, Bohr's institute soon accomplished these goals to the highest degree.

In his 1913 trilogy, Bohr had sought to apply his theory to the understanding of the periodic table of elements. He improved upon this aspect of his work into the early 1920s, by which time he had developed an elaborate scheme building up the periodic table by adding electrons one after another to the atom according to his atomic model. When Bohr was awarded the Nobel Prize for his work in 1922, the Hungarian physical chemist Georg Hevesy, together with the physicist Dirk Coster from Holland, were working at Bohr's institute to establish experimentally that the as-yet-undiscovered atomic element 72 would behave as predicted by Bohr's theory. They succeeded in 1923, thus proving both the strength of Bohr's theory and the truth in practice of Bohr's words at the institute's inauguration about the important role of experiment. The element was named hafnium (Latin for Copenhagen).

Among physicists working at Bohr's institute between the World Wars, the "Copenhagen Spirit" came to denote the very special social milieu there, comprising a completely informal atmosphere, the opportunity to discuss physics without any concern for other matters, and, for the specially privileged, the unique opportunity of working with Bohr.

Notwithstanding the important experimental work performed by Hevesy, Coster, and others, it was the theorists who led the way. In 1925 Werner Heisenberg of Germany developed the revolutionary quantum mechanics, which, in contrast to its predecessor, the so-called "old quantum theory" that drew on classical physics, constituted a fully independent theory. During the academic year 1926–27, Heisenberg served as Bohr's assistant in Copenhagen, where he formulated the fundamental uncertainty principle as a consequence of quantum mechanics. Bohr, Heisenberg, and a few others then went on to develop what came to be known as the Copenhagen interpretation of quantum mechanics, which still provides a conceptual basis for the theory. A central element of the Copenhagen interpretation is Bohr's complementarity principle, presented for the first time in 1927 at a conference in Como, Italy. According to complementarity, on the atomic level a physical phenomenon expresses itself differently depending on the experimental setup used to observe it. Thus, light appears sometimes as waves and sometimes as particles. For a complete explanation, both aspects, which according to classical physics are contradictory, need to be taken into account. The other towering figure of physics in the 20th century, Albert Einstein, never accepted the Copenhagen interpretation, famously declaring against its probabilistic implications that "God does not play dice." The discussions between Bohr and Einstein, especially at two of the renowned series of Solvay Conferences in physics, in 1927 and 1930, constitute one of the most fundamental and inspired discussions between physicists in the 20th century. For the rest of his life, Bohr worked to generalize complementarity as a guiding idea applying far beyond physics.

In the early 1930s Bohr found use once more for his fundraising abilities and his vision of a fruitful combination of theory and experiment. He realized early that the research front in theoretical physics was moving from the study of the atom as a whole to the study of its nucleus. Bohr turned to the Rockefeller Foundation, whose "experimental biology" program was designed to improve conditions for the life sciences. Together with Hevesy and the Danish physiologist August Krogh, Bohr applied for support to build a cyclotron—a kind of particle accelerator recently invented by Ernest O. Lawrence in the United States—as a means to pursue biological studies. Although Bohr intended to use the cyclotron primarily for investigations in nuclear physics, it could also produce isotopes of elements involved in organic processes, making it possible in particular to extend the radioactive indicator method, invented and promoted by Hevesy, to biological purposes. In addition to the support from the Rockefeller Foundation, funds for the cyclotron and other equipment for studying the nucleus were also granted to Bohr from Danish sources.

Just as the close connection between theory and experiment had proved fruitful for atomic physics, so now the same connection came to work well in the study of the nucleus. Thus, after the German physicists Otto Hahn and Fritz Strassmann in late 1938 had made the unexpected and unexplained experimental discovery that a uranium atom can be split in two approximately equal halves when bombarded with neutrons, a theoretical explanation based on Bohr's recently proposed theory of the compound nucleus was suggested by two Austrian physicists close to Bohr—Lise Meitner and her nephew Otto Robert Frisch; the explanation was soon confirmed in experiments by Meitner and Frisch at the institute. By this time, at the beginning of 1939, Bohr was in the United States, where a fierce race to confirm experimentally the so-called fission of the nucleus began after the news of the German experiments and their explanation had become known. In the United States, Bohr did pathbreaking work with his younger American colleague John Archibald Wheeler at Princeton University to explain fission theoretically.

Bohr had felt the consequences of the Nazi regime almost as soon as Adolf Hitler came to power in Germany in 1933, as several of his colleagues there were of Jewish descent and lost their jobs without any prospect of a future in their home country. Bohr used his connections with well-established foundations—as well as the newly set up Danish Committee for the Support of Refugee Intellectual Workers, in which he sat on the executive board from its creation in 1933—to get physicists out of Germany in order for them to spend some time at Bohr's institute before obtaining permanent appointment elsewhere, most often in the United States.

After the discovery of fission, Bohr was acutely aware of the theoretical possibility of making an atomic bomb. However, as he announced in lectures in Denmark and in Norway just before the German occupation of both countries in April 1940, he considered the practical difficulties so prohibitive as to prevent the realization of a bomb until well after the war could be expected to end. Even when Heisenberg at his visit to Copenhagen in 1941 told Bohr about his role in a German atomic bomb project, Bohr did not waver from this conviction.

In early 1943 Bohr received a secret message from his British colleague James Chadwick, inviting Bohr to join him in England to do important scientific work. Although Chadwick's letter was vaguely formulated, Bohr understood immediately that the work had to do with developing an atomic bomb. Still convinced of the infeasibility of such a project, Bohr answered that there was greater need for him in occupied Denmark.

In the fall of 1943, the political situation in Denmark changed dramatically after the Danish government's collaboration with the German occupiers broke down. After being warned about his imminent arrest, Bohr escaped by boat with his family across the narrow sound to Sweden. In Stockholm the invitation to England was repeated, and Bohr was brought by a military airplane to Scotland and then on to London. Only a few days later he was joined by his son Aage, a fledgling physicist of age 21, who would serve as his father's indispensable sounding board during their absence from Denmark.

Upon being briefed about the state of the Allied atomic bomb project on his arrival in London, Bohr changed his mind immediately about its feasibility. Concerned about a corresponding project being pursued in Germany, Bohr willingly joined the Allied project. Taking part for several weeks at a time in the work in Los Alamos, N.M., to develop the atomic bomb, he made significant technical contributions, notably to the design of the so-called initiator for the plutonium bomb. His most important role, however, was to serve, in J. Robert Oppenheimer's words, "as a scientific father confessor to the younger men."

Early on during his exile, Bohr became convinced that the existence of the bomb would "not only seem to necessitate but should also, due to the urgency of mutual confidence, facilitate a new approach to the problems of international relationship." The first step toward avoiding a postwar nuclear arms race would be to inform the ally in the war, the Soviet Union, of the project. Bohr set out on a solitary campaign, during which he even succeeded in obtaining personal interviews with British Prime Minister Winston Churchill and U.S. President Franklin D. Roosevelt. He was unable to convince either of them of his viewpoint, however, instead being suspected by Churchill of spying for the Russians. After the war, Bohr persisted in his mission for what he called an "open world" between nations, continuing his confidential contact with statesmen and writing an open letter to the United Nations in 1950.

Bohr was allowed to return home only after the atomic bomb had been dropped on Japan in August 1945. In Denmark he was greeted as a hero some newspapers even welcoming him with pride as the Dane who had invented the atomic bomb. He continued to run and expand his institute, and he was central in postwar institution building for physics. On a national scale, he took a major part in establishing the research facility at Risø, near Roskilde, only a few miles outside Copenhagen, created in order to prepare the introduction of nuclear power in Denmark, which, however, has never occurred. Internationally, he took part in the establishment of CERN, the European experimental particle physics facility near Geneva, Switz., as well as of the Nordic Institute for Atomic Physics (Nordita) adjacent to his institute. Bohr left behind an unsurpassed scientific legacy, as well as an institute that remains one of the leading centres for theoretical physics in the world.

MAX BORN

(b. Dec. 11, 1882, Breslau, Ger. [now Wrocław, Pol.]—d. Jan. 5, 1970, Göttingen, W.Ger.)

German physicist Max Born shared the Nobel Prize for Physics in 1954 with Walther Bothe for his probabilistic interpretation of quantum mechanics.

Born came from an upper-middle-class, assimilated Jewish family. At first he was considered too frail to attend public school, so he was tutored at home before being allowed to attend the König Wilhelm Gymnasium in Breslau. Thereafter he continued his studies in physics

Max Born was a German physicist who advanced the development of quantum mechanics. Keystone/Hulton Archive/Getty Images

and mathematics at universities in Breslau, Heidelberg, Zürich, and Göttingen. At the University of Göttingen he wrote his dissertation (1906) on the stability of elastic wires and tapes under the direction of the mathematician Felix Klein, for which he was awarded a doctorate in 1907.

After brief service in the army and a stay at the University of Cambridge, where he worked with physicists Joseph Larmor and J.J. Thomson, Born returned to Breslau for the academic year 1908–09 and began an extensive study of Albert Einstein's theory of special relativity. On the strength of his papers in this field, Born was invited back to Göttingen as an assistant to the mathematical physicist Hermann Minkowski. In 1912 Born met Hedwig Ehrenberg, whom he married a year later. Three children, two girls and a boy, were born from the union. It was a troubled relationship, and Born and his wife often lived apart.

In 1915 Born accepted a professorship to assist physicist Max Planck at the University of Berlin, but World War I intervened and he was drafted into the German army. Nonetheless, while an officer in the army, he found time to publish his first book, *Dynamik der Kristallgitter* (1915; *Dynamics of Crystal Lattices*).

In 1919 Born was appointed to a full professorship at the University of Frankfurt am Main, and in 1921 he accepted the position of professor of theoretical physics at the University of Göttingen. James Franck had been appointed professor of experimental physics at Göttingen the previous year. The two of them made the University of Göttingen one of the most important centres for the study of atomic and molecular phenomena. A measure of Born's influence can be gauged by the students and assistants who came to work with him—among them, Wolfgang Pauli, Werner Heisenberg, Pascual Jordan, Enrico Fermi, Fritz London, P.A.M. Dirac, Victor Weisskopf, J. Robert Oppenheimer, Walter Heitler, and Maria Goeppert-Mayer.

The Göttingen years were Born's most creative and seminal. In 1912 Born and Hungarian engineer Theodore von Kármán formulated the dynamics of a crystal lattice, which incorporated the symmetry properties of the lattice, allowed the imposition of quantum rules, and permitted thermal properties of the crystal to be calculated. This work was elaborated when Born was in Göttingen, and it formed the basis of the modern theory of lattice dynamics.

In 1925 Heisenberg gave Born a copy of the manuscript of his first paper on quantum mechanics, and Born immediately recognized that the mathematical entities with which Heisenberg had represented the observable physical quantities of a particle—such as its position, momentum, and energy—were matrices. Joined by Heisenberg and Jordan, Born formulated all the essential aspects of quantum mechanics in its matrix version. A short time later, Erwin Schrödinger formulated a version of quantum mechanics based on his wave equation. It was soon proved that the two formulations were mathematically equivalent. What remained unclear was the meaning of the wave function that appeared in Schrödinger's equation. In 1926 Born submitted two papers in which he formulated the quantum mechanical description of collision processes and found that in the case of the scattering of a particle by a potential, the wave function at a particular spatiotemporal location should be interpreted as the probability amplitude of finding the particle at that specific space-time point. In 1954 he was awarded the Nobel Prize for this work.

Born remained at Göttingen until April 1933, when all Jews were dismissed from their academic posts in Germany. Born and his family went to England, where he accepted a temporary lectureship at Cambridge. In 1936 he was appointed Tait Professor of Natural Philosophy at the University of Edinburgh. He became a British citizen in 1939 and remained at Edinburgh until his retirement in 1953. The next year, he and his wife moved to Bad Pyrmont, a small spa town near Göttingen.

SATYENDRA NATH BOSE

(b. Jan. 1, 1894, Calcutta, India—d. Feb. 4, 1974, Calcutta)

Indian mathematician and physicist Satyendra Nath Bose was noted for his collaboration with Albert Einstein in developing a theory regarding the gaslike qualities of electromagnetic radiation.

Bose, a graduate of the University of Calcutta, taught at the University of Dacca (1921–45) and then at Calcutta (1945–56). Bose's numerous scientific papers (published from 1918 to 1956) contributed to statistical mechanics, the electromagnetic properties of the ionosphere, the theories of X-ray crystallography and thermoluminescence, and unified field theory. Bose's *Planck's Law and the Hypothesis of Light Quanta* (1924) led Einstein to seek him out for collaboration.

LOUIS-VICTOR, 7^E DUKE DE BROGLIE

(b. Aug. 15, 1892, Dieppe, France—d. March 19, 1987, Paris)

French physicist Louis-Victor, the seventh duke de Broglie, was best known for his research on quantum theory and for his discovery of the wave nature of electrons. He was awarded the 1929 Nobel Prize for Physics.

Broglie was the second son of a member of the French nobility. From the Broglie family, whose name is taken from a small town in Normandy, have come high-ranking soldiers, politicians, and diplomats since the 17th century.

In choosing science as a profession, Louis de Broglie broke with family tradition, as had his brother Maurice (from whom, after his death, Louis inherited the title of duke). Maurice, who was also a physicist and made notable contributions to the experimental study of the atomic nucleus, kept a well-equipped laboratory in the family mansion in Paris. Louis occasionally joined his brother in his work, but it was the purely conceptual side of physics that attracted him. He described himself as "having much more the state of mind of a pure theoretician than that of an experimenter or engineer, loving especially the general and philosophical view...." He was brought into one of his few contacts with the technical aspects of physics during World War I, when he saw army service in a radio station in the Eiffel Tower.

Broglie's interest in what he called the "mysteries" of atomic physics—namely, unsolved conceptual problems of the science—was aroused when he learned from his brother about the work of the German physicists Max Planck and Albert Einstein, but the decision to take up the profession of physicist was long in coming. He began at 18 to study theoretical physics at the Sorbonne, but he was also earning his degree in history (1909), thus moving along the family path toward a career in the diplomatic service. After a period of severe conflict, he declined the research project in French history that he had been assigned and chose for his doctoral thesis a subject in physics.

In this thesis (1924) Broglie developed his revolutionary theory of electron waves, which he had published earlier in scientific journals. The notion that matter on the atomic scale might have the properties of a wave was rooted in a proposal Einstein had made 20 years before. Einstein had suggested that light of short wavelengths might under some conditions be observed to behave as if it were composed of particles, an idea that was confirmed in 1923. The dual nature of light, however, was just beginning to gain scientific acceptance when Broglie extended the idea of such a duality to matter.

Broglie's proposal answered a question that had been raised by calculations of the motion of electrons within the atom. Experiments had indicated that the electron must move around a nucleus and that, for reasons then obscure, there are restrictions on its motion. Broglie's idea of an electron with the properties of a wave offered an explanation of the restricted motion. A wave confined within boundaries imposed by the nuclear charge would be restricted in shape and, thus, in motion, for any wave shape that did not fit within the atomic boundaries would interfere with itself and be canceled out. In 1923, when Broglie put forward this idea, there was no experimental evidence whatsoever that the electron, the corpuscular properties of which were well established by experiment, might under some conditions behave as if it were radiant energy. Broglie's suggestion, his one major contribution to physics, thus constituted a triumph of intuition.

The first publications of Broglie's idea of "matter waves" had drawn little attention from other physicists, but a copy of his doctoral thesis chanced to reach Einstein, whose response was enthusiastic. Einstein stressed the importance of Broglie's work both explicitly and by building further on it. In this way the Austrian physicist Erwin Schrödinger learned of the hypothetical waves, and on the basis of the idea, he constructed a mathematical system, wave mechanics, that has become an essential tool of physics. Not until 1927, however, did Clinton Davisson and Lester Germer in the United States and George Thomson in Scotland find the first experimental evidence of the electron's wave nature.

After receiving his doctorate, Broglie remained at the Sorbonne, becoming in 1928 professor of theoretical physics at the newly founded Henri Poincaré Institute, where he taught until his retirement in 1962. He also acted, after 1945, as an adviser to the French Atomic Energy Commissariat.

In addition to winning the Nobel Prize for Physics, Broglie received, in 1952, the Kalinga Prize, awarded by the United Nations Economic and Social Council, in recognition of his writings on science for the general public. He was a foreign member of the British Royal Society, a member of the French Academy of Sciences, and, like several of his forebears, a member of the Académie Française.

Broglie's keen interest in the philosophical implications of modern physics found expression in addresses, articles, and books. The central question for him was whether the statistical considerations that are fundamental to atomic physics reflect an ignorance of underlying causes or whether they express all that there is to be known; the latter would be the case if, as some believe, the act of measuring affects, and is inseparable from, what is measured. For about three decades after his work of 1923, Broglie held the view that underlying causes could not be delineated in a final sense, but, with the passing of time, he returned to his earlier belief that the statistical theories hide "a completely determined and ascertainable reality behind variables which elude our experimental techniques."

EDWARD UHLER CONDON

(b. March 2, 1902, Alamogordo, N.M., U.S.—d. March 26, 1974, Boulder, Colo.)

Edward Uhler Condon was an American physicist for whom the Franck–Condon principle was named and who

Edward U. Condon. National Institute of Standards and Technology

applied quantum mechanics to an understanding of the atom and its nucleus.

During World War II Condon made valuable contributions to the development of both atomic energy and radar. In 1943 he helped J. Robert Oppenheimer recruit the group that made the first atomic bombs at Los Alamos, N.M. In 1946 he was a consultant to the committee of the Senate that drafted the legislation that created the Atomic Energy Commission; in the aftermath of the struggle to put atomic energy under civilian control he was attacked by the House Committee on Un-American Activities, one of the strongest opponents of civilian control. Condon was director of the National Bureau of Standards (1945–51) and president of both the American Physical Society (1946) and the American Association for the Advancement of Science (1953). In 1966 the Air Force Office of Scientific Research appointed him director of a project to investigate flying saucers, from which grew the Condon report, *The Scientific Study of Unidentified Flying Objects* (1969).

CLINTON JOSEPH DAVISSON

(b. Oct. 22, 1881, Bloomington, Ill., U.S.—d. Feb. 1, 1958, Charlottesville, Va.)

American experimental physicist Clinton Joseph Davisson shared the Nobel Prize for Physics in 1937 with George P. Thomson of England for discovering that electrons can be diffracted like light waves, thus verifying the thesis of Louis de Broglie that electrons behave both as waves and as particles.

Davisson received his doctorate from Princeton University and spent most of his career at the Bell Telephone Laboratories. He began his research there on the emissions of electrons from a metal in the presence of heat and later helped develop the electron microscope.

Then, in 1927, Davisson and Lester H. Germer found that a beam of electrons, when reflected from a metallic crystal, shows diffraction patterns similar to those of X-rays and other electromagnetic waves. This discovery verified quantum mechanics' understanding of the dual nature of subatomic particles and proved to be useful in the study of nuclear, atomic, and molecular structure.

P.A.M. DIRAC

(b. Aug. 8, 1902, Bristol, Gloucestershire, Eng.—d. Oct. 20, 1984, Tallahassee, Fla., U.S.)

Paul Adrien Maurice Dirac was an English theoretical physicist who was one of the founders of quantum mechanics and quantum electrodynamics. Dirac is most famous for his 1928 relativistic quantum theory of the electron and his prediction of the existence of antiparticles. In 1933 he shared the Nobel Prize for Physics with the Austrian physicist Erwin Schrödinger.

Dirac's mother was British and his father was Swiss. Dirac's childhood was not happy—his father intimidated the children, both at home and at school where he taught French, by meticulous and oppressive discipline. Dirac grew up an introvert, spoke only when spoken to, and used words very sparingly—though with utmost precision in meaning. In later life, Dirac would become proverbial for his lack of social and emotional skills and his incapacity for small talk. He preferred solitary thought and long walks to company and had few, though very close, friends. Dirac showed from early on extraordinary mathematical abilities but hardly any interest in literature and art. His physics papers and books, however, are literary masterpieces of the genre owing to their absolute perfection in form with regard to mathematical expressions as well as words.

On his father's wish for a practical profession for his sons, Dirac studied electrical engineering at the University of Bristol (1918–21). Having not found employment upon graduation, he took two more years of applied mathematics. Albert Einstein's theory of relativity had become famous after 1919 through the mass media. Fascinated with the technical aspect of relativity, Dirac mastered it on his own. Following the advice of his mathematics professors, and with the help of a fellowship, he entered the University of Cambridge as a research student in 1923. Dirac had no teacher in the true sense, but his adviser, Ralph Fowler, was then the only professor in Cambridge at home with the new quantum theory being developed in Germany and Denmark.

In August 1925 Dirac received through Fowler proofs of an unpublished paper by Werner Heisenberg that initiated the revolutionary transition from the Bohr atomic model to the new quantum mechanics. In a series of papers and his 1926 Ph.D. thesis, Dirac further developed Heisenberg's ideas. Dirac's accomplishment was more general in form but similar in results to matrix mechanics, another early version of quantum mechanics created about the same time in Germany by a joint effort of Heisenberg, Max Born, Pascual Jordan, and Wolfgang Pauli. In the fall of 1926 Dirac and, independently, Jordan combined the matrix approach with the powerful methods of Schrödinger's wave mechanics and Born's statistical interpretation into a general scheme—transformation theory—that was the first complete mathematical formalism of quantum mechanics. Along the way, Dirac also developed the Fermi-Dirac statistics (which had been suggested somewhat earlier by Enrico Fermi).

Satisfied with the interpretation that the fundamental laws governing microscopic particles are probabilistic, or that "nature makes a choice," Dirac declared quantum mechanics complete and turned his main attention to relativistic quantum theory. Often regarded as the true beginning of quantum electrodynamics is his 1927 quantum theory of radiation. In it Dirac developed methods of quantizing electromagnetic waves and invented the so-called second quantization—a way to transform the description of a single quantum particle into a formalism of the system of many such particles. In 1928 Dirac published what may be his greatest single accomplishment—the relativistic wave equation for the electron. In order to satisfy the condition of relativistic invariance (i.e., treating space and time coordinates on the same footing), the Dirac equation required a combination of four wave functions and relatively new mathematical quantities known as spinors.

As an added bonus, the equation described electron spin (magnetic moment)—a fundamental but theretofore not properly explained feature of quantum particles.

From the beginning, Dirac was aware that his spectacular achievement also suffered grave problems: it had an extra set of solutions that made no physical sense, as it corresponded to negative values of energy. In 1930 Dirac suggested a change in perspective to consider unoccupied vacancies in the sea of negative-energy electrons as positively charged "holes." By suggesting that such "holes" could be identified with protons, he hoped to produce a unified theory of matter, as electrons and protons were then the only known elementary particles. Others proved, however, that a "hole" must have the same mass as the electron, whereas the proton is a thousand times heavier. This led Dirac to admit in 1931 that his theory, if true, implied the existence of "a new kind of particle, unknown to experimental physics, having the same mass and opposite charge to an electron." One year later, to the astonishment of physicists, this particle—the antielectron, or positron—was accidentally discovered in cosmic rays by Carl Anderson of the United States.

An apparent difficulty of the Dirac equation thus turned into an unexpected triumph and one of the main reasons for Dirac's being awarded the 1933 Nobel Prize for Physics. The power to predict unexpected natural phenomena is often the most convincing argument in favour of novel theories. In this regard the positron of quantum theory has often been compared to the planet Neptune, the discovery of which in the 19th century was spectacular proof of the astronomical precision and predictive power of classical Newtonian science. Dirac drew from this experience a methodological lesson that theoretical physicists, in their quest for new laws, should place more trust in mathematical formalism and follow its lead, even if physical understanding of the formulas temporarily lags behind. In later life, he often expressed the view that, in order to be true, a fundamental physical theory must also be mathematically beautiful. Dirac's prediction of another new particle in 1931—the magnetic monopole—seems to have demonstrated that mathematical beauty is a necessary but not sufficient condition for physical truth, as no such particle has been discovered. Numerous other elementary particles discovered after 1932 by experimental physicists were, more often than not, stranger and messier than anything the theorists could have anticipated on the basis of mathematical formulas. But for each of these new particles, an antiparticle also exists—a universal property of matter first uncovered by Dirac.

In his later work, Dirac continued making important improvements and clarifications in the logical and mathematical presentation of quantum mechanics, in particular through his influential textbook *The Principles of Quantum Mechanics* (1930, with three subsequent major revisions). The professional terminology of modern theoretical physics owes much to Dirac, including the names and mathematical notations *fermion*, *boson*, *observable*, *commutator*, *eigenfunction*, *delta-function*, $\hbar$ (for $h/2\pi$, where h is Planck's constant), and the bra-ket vector notation.

Compared with the standard of logical clarity that Dirac accomplished in his formalization of quantum mechanics, relativistic quantum theory seemed incomplete to him. In the 1930s quantum electrodynamics encountered serious problems; in particular, infinite results appeared in various mathematical calculations. Dirac was even more concerned with the formal difficulty that relativistic invariance did not follow directly from the main equations, which treated time and space coordinates separately. Searching for remedies, Dirac in 1932–33 introduced the "many-times formulation" (sometimes called "interaction representation") and the quantum analog for the principle of least action, later developed by Richard Feynman into the method of path integration. These concepts, and also Dirac's idea of vacuum polarization (1934), helped a new generation of theorists after World War II invent ways of subtracting infinities from one another in their calculations so that predictions for physically observable results in quantum electrodynamics would always be finite quantities. Although very effective in practical calculations, these "renormalization" techniques remained, in Dirac's view, clever tricks rather than a principled solution to a fundamental problem. He hoped for a revolutionary change in basic principles that would eventually bring the theory to a degree of logical consistency comparable to what had been achieved in nonrelativistic quantum mechanics. Although Dirac probably contributed more to quantum electrodynamics than any other physicist, he died dissatisfied with his own brainchild.

Dirac taught at Cambridge after receiving his doctorate there, and in 1932 he was appointed Lucasian Professor of Mathematics, the chair once held by Isaac Newton. Although Dirac had few research students, he was very active in the research community through his participation in international seminars. Unlike many physicists of his generation and expertise, Dirac did not switch to nuclear physics and only marginally participated in the development of the atomic bomb during World War II. In 1937 he married Margit Balasz (née Wigner; sister of Hungarian physicist Eugene Wigner). Dirac retired from Cambridge in 1969 and, after various visiting appointments, held a professorship at Florida State University, Tallahassee, from 1971 until his death.

SIR ARTHUR STANLEY EDDINGTON

(b. Dec. 28, 1882, Kendal, Westmorland, Eng.—d. Nov. 22, 1944, Cambridge, Cambridgeshire)

English astronomer, physicist, and mathematician Sir Arthur Stanley Eddington did his greatest work in astrophysics, investigating the motion, internal structure, and evolution of stars. He also was the first expositor of the theory of relativity in the English language.

Eddington was the son of the headmaster of Stramongate School, an old Quaker foundation in Kendal near Lake Windermere in the northwest of England. His father, a gifted and highly educated man, died of typhoid in 1884. The widow took her daughter and small son to Weston-super-Mare in Somerset, where young Eddington grew up and received his schooling. He entered Owens College, Manchester, in October 1898, and Trinity College, Cambridge, in October 1902. There he won every mathematical honour, as well as Senior Wrangler (1904), Smith's prize, and a Trinity College fellowship (1907). In 1913 he received the Plumian Professorship of Astronomy at Cambridge and in 1914 became also the director of its observatory.

From 1906 to 1913 Eddington was chief assistant at the Royal Observatory at Greenwich, where he gained practical experience in the use of astronomical instruments. He made observations on the island of Malta to establish its longitude, led an eclipse expedition to Brazil, and investigated the distribution and motions of the stars. He broke new ground with a paper on the dynamics of a globular stellar system. In *Stellar Movements and the Structure of the Universe* (1914) he summarized his mathematically elegant investigations, putting forward the thesis that the spiral nebulae, cloudy structures seen in the telescope, were galaxies like the Milky Way.

During World War I he declared himself a pacifist. This arose out of his strongly held Quaker beliefs. His religious faith also found expression in his popular writings on the philosophy of science. In *Science and the Unseen World* (1929) he declared that the world's meaning could not be discovered from science but must be sought through apprehension of spiritual reality. He expressed this belief in other philosophical books: *The Nature of the Physical World* (1928), *New Pathways of Science* (1935), and *The Philosophy of Physical Science* (1939).

During these years he carried on important studies in astrophysics and relativity, in addition to teaching and lecturing. In 1919 he led an expedition to Príncipe Island (West Africa) that provided the first confirmation of Einstein's theory that gravity will bend the path of light when it passes near a massive star. During the total eclipse of the sun, it was found that the positions of stars seen just beyond the eclipsed solar disk were, as the general theory of relativity had predicted, slightly displaced away from the centre of the solar disk. His *Report on the Relativity Theory of Gravitation* (1918), written for the Physical Society, followed by *Space, Time and Gravitation* (1920) and his great treatise *The Mathematical Theory of Relativity* (1923)—the latter considered by Einstein the finest presentation of the subject in any language—made Eddington a leader in the field of relativity physics. His own contribution was chiefly a brilliant modification of affine (non-Euclidean) geometry, leading to a geometry of the cosmos. Later, when the Belgian astronomer Georges Lemaître produced the hypothesis of the expanding universe, Eddington pursued the subject in his own researches; these were placed before the general reader in his little book *The Expanding Universe* (1933). Another book, *Relativity Theory of Protons and Electrons* (1936), dealt with quantum theory. He gave many popular lectures on relativity, leading the English physicist Sir Joseph John Thomson to remark that Eddington had persuaded multitudes of people that they understood what relativity meant.

His philosophical ideas led him to believe that through a unification of quantum theory and general relativity it would be possible to calculate the values of universal constants, notably the fine-structure constant, the ratio of the mass of the proton to that of the electron, and the number of atoms in the universe. This was an attempt, never completed, at a vast synthesis of the known facts of the physical universe; it was published posthumously as *Fundamental Theory* (1946), edited by Sir Edmund Taylor Whittaker, a book that is incomprehensible to most readers and perplexing in many places to all, but which represents a continuing challenge to some.

Eddington received many honours, including honorary degrees from 12 universities. He was president of the Royal Astronomical Society (1921–23), the Physical Society (1930–32), the Mathematical Association (1932), and the International Astronomical Union (1938–44). He was knighted in 1930 and received the Order of Merit in 1938. Meetings of the Royal Astronomical Society were often enlivened by dramatic clashes between Eddington and Sir James Hopwood Jeans or Edward Arthur Milne over the validity of scientific assumptions and mathematical procedures. Eddington was an enthusiastic participant in most forms of athletics, confining himself in later years to cycling, swimming, and golf.

Eddington's greatest contributions were in the field of astrophysics, where he did pioneer work on stellar structure and radiation pressure, subatomic sources of stellar energy, stellar diameters, the dynamics of pulsating stars, the relation between stellar mass and luminosity, white dwarf stars, diffuse matter in interstellar space, and so-called forbidden spectral lines. His work in astrophysics is represented by the classic *Internal Constitution of the Stars* (1925) and in the public lectures published as *Stars and Atoms* (1927). In his well-written popular books he also set forth his scientific epistemology, which he called "selective subjectivism" and "structuralism"—i.e., the interplay of physical observations and geometry. He believed that a great part of physics simply reflected the interpretation that the scientist imposes on his data. The better part of his philosophy, however, was not his metaphysics but his "structure" logic. His theoretical work in physics had a stimulating effect on the thought and research of others, and many lines of scientific investigation were opened as a result of his work.

ALBERT EINSTEIN

(b. March 14, 1879, Ulm, Württemberg, Ger.—d. April 18, 1955, Princeton, N.J., U.S.)

German-born physicist Albert Einstein developed the special and general theories of relativity and won the Nobel Prize for Physics in 1921 for his explanation of the photoelectric effect. Einstein is generally considered the most influential physicist of the 20th century.

Einstein's parents were secular, middle-class Jews. His father, Hermann Einstein, was originally a featherbed salesman and later ran an electrochemical factory with moderate success. His mother, the former Pauline Koch, ran the family household. He had one sister, Maja, born two years after Albert.

Einstein would write that two "wonders" deeply affected his early years. The first was his encounter with a compass at age five. He was mystified that invisible forces could deflect the needle. This would lead to a lifelong fascination with invisible forces. The second wonder came at age 12 when he discovered a book of geometry, which he devoured, calling it his "sacred little geometry book."

Einstein became deeply religious at age 12, even composing several songs in praise of God and chanting religious songs on the way to school. This began to change, however, after he read science books that contradicted his religious beliefs. This challenge to established authority left a deep and lasting impression. At the Luitpold Gymnasium, Einstein often felt out of place and victimized by a Prussian-style educational system that seemed to stifle originality and creativity. One teacher even told him that he would never amount to anything.

Yet another important influence on Einstein was a young medical student, Max Talmud (later Max Talmey), who often had dinner at the Einstein home. Talmud became an informal tutor, introducing Einstein to higher mathematics and philosophy. A pivotal turning point occurred when Einstein was 16. Talmud had earlier introduced him to a children's science series by Aaron Bernstein, *Naturwissenschaftliche Volksbucher* (1867–68; *Popular Books on Physical Science*), in which the author imagined riding alongside electricity that was traveling inside a telegraph wire. Einstein then asked himself the question that would dominate his thinking for the next 10 years: What would a light beam look like if you could run alongside it? If light were a wave, then the light beam should appear stationary, like a frozen wave. Even as a child, though, he knew that stationary light waves had never been seen, so there was a paradox. Einstein also wrote his first "scientific paper" at that time ("The Investigation of the State of Aether in Magnetic Fields").

Einstein's education was disrupted by his father's repeated failures at business. In 1894, after his company failed to get an important contract to electrify the city of Munich, Hermann Einstein moved to Milan, Italy, to work with a relative. Einstein was left at a boarding house in Munich and expected to finish his education. Alone, miserable, and repelled by the looming prospect of military duty when he turned 16, Einstein ran away six months later and landed on the doorstep of his surprised parents. His parents realized the enormous problems that he faced as a school dropout and draft dodger with no employable skills. His prospects did not look promising.

Fortunately, Einstein could apply directly to the Eidgenössische Polytechnische Schule ("Swiss Federal Polytechnic School"; in 1911, following expansion in 1909 to full university status, it was renamed the Eidgenössische Technische Hochschule, or "Swiss Federal Institute of Technology") in Zürich without the equivalent of a high school diploma if he passed its stiff entrance examinations. His marks showed that he excelled in mathematics and physics, but he failed at French, chemistry, and biology. Because of his exceptional math scores, he was allowed into the polytechnic on the condition that he first finish his formal schooling. He went to a special high school run by Jost Winteler in Aarau, Switz., and graduated in 1896. He also renounced his German citizenship at that time. (He was stateless until 1901, when he was granted Swiss citizenship.) He became lifelong friends with the Winteler family, with whom he had been boarding. (Winteler's daughter, Marie, was Einstein's first love; Einstein's sister Maja would eventually marry Winteler's son Paul; and his close friend Michele Besso would marry their eldest daughter, Anna.)

Einstein would recall that his years in Zürich were some of the happiest years of his life. He met many students who would become loyal friends, such as Marcel Grossmann, a mathematician, and Besso, with whom he enjoyed lengthy conversations about space and time. He also met his future wife, Mileva Maric, a fellow physics student from Serbia.

After graduation in 1900, Einstein faced one of the greatest crises in his life. Because he studied advanced subjects on his own, he often cut classes; this earned him the animosity of some professors, especially Heinrich Weber. Unfortunately, Einstein asked Weber for a letter of recommendation. Einstein was subsequently turned down for every academic position that he applied to. He later wrote,

> *I would have found [a job] long ago if Weber had not played a dishonest game with me.*

Meanwhile, Einstein's relationship with Maric deepened, but his parents vehemently opposed the relationship. His mother especially objected to her Serbian background (Maric's family was Eastern Orthodox Christian). Einstein defied his parents, however, and he and Maric even had a child, Lieserl, in January 1902, whose fate is unknown. (It is commonly thought that she died of scarlet fever or was given up for adoption.)

In 1902 Einstein reached perhaps the lowest point in his life. He could not marry Maric and support a family without a job, and his father's business went bankrupt. Desperate and unemployed, Einstein took lowly jobs tutoring children, but he was fired from even these jobs.

The turning point came later that year, when the father of his lifelong friend, Marcel Grossman, was able to recommend him for a position as a clerk in the Swiss patent office in Bern. About then Einstein's father became seriously ill and, just before he died, gave his blessing for his son to marry Maric. For years, Einstein would experience enormous sadness remembering that his father had died thinking him a failure.

With a small but steady income for the first time, Einstein felt confident enough to marry Maric, which he did on Jan. 6, 1903. Their children, Hans Albert and Eduard, were born in Bern in 1904 and 1910, respectively. In hindsight, Einstein's job at the patent office was a blessing. He would quickly finish analyzing patent applications, leaving him time to daydream about the vision that had obsessed him since he was 16: What will happen if you race alongside a light beam? While at the polytechnic school he had studied Maxwell's equations, which describe the nature of light, and discovered a fact unknown to James Clerk Maxwell himself—namely, that the speed of light remained the same no matter how fast one moved. This violated Newton's laws of motion, however, because there is no absolute velocity in Isaac Newton's theory. This insight led Einstein to formulate the principle of relativity: "the speed of light is a constant in any inertial frame (constantly moving frame)."

During 1905, often called Einstein's "miracle year," he published four papers in the *Annalen der Physik*, each of which would alter the course of modern physics:

1. "Über einen die Erzeugung und Verwandlung des Lichtes betreffenden heuristischen Gesichtspunkt" ("On a Heuristic Viewpoint Concerning the Production and Transformation of Light"), in which Einstein applied the quantum theory to light in order to explain the photoelectric effect. If light occurs in tiny packets (later called photons), then it should knock out electrons in a metal in a precise way.
2. "Über die von der molekularkinetischen Theorie der Wärme geforderte Bewegung

von in ruhenden Flüssigkeiten suspendierten Teilchen" ("On the Movement of Small Particles Suspended in Stationary Liquids Required by the Molecular-Kinetic Theory of Heat"), in which Einstein offered the first experimental proof of the existence of atoms. By analyzing the motion of tiny particles suspended in still water, called Brownian motion, he could calculate the size of the jostling atoms and Avogadro's number.

3. "Zur Elektrodynamik bewegter Körper" ("On the Electrodynamics of Moving Bodies"), in which Einstein laid out the mathematical theory of special relativity.

4. "Ist die Trägheit eines Körpers von seinem Energieinhalt abhängig?" ("Does the Inertia of a Body Depend Upon Its Energy Content?"), submitted almost as an afterthought, which showed that relativity theory led to the equation $E = mc^2$. This provided the first mechanism to explain the energy source of the Sun and other stars.

Einstein also submitted a paper in 1905 for his doctorate.

Other scientists, especially Henri Poincaré and Hendrik Lorentz, had pieces of the theory of special relativity, but Einstein was the first to assemble the whole theory together and to realize that it was a universal law of nature, not a curious figment of motion in the ether, as Poincaré and Lorentz had thought. (In one private letter to Mileva, Einstein referred to "our theory," which has led some to speculate that she was a cofounder of relativity theory. However, Mileva had abandoned physics after twice failing her graduate exams, and there is no record of her involvement in developing relativity. In fact, in his

1905 paper, Einstein only credits his conversations with Besso in developing relativity.)

In the 19th century there were two pillars of physics: Newton's laws of motion and Maxwell's theory of light. Einstein was alone in realizing that they were in contradiction and that one of them must fall.

At first Einstein's 1905 papers were ignored by the physics community. This began to change after he received the attention of just one physicist, perhaps the most influential physicist of his generation, Max Planck, the founder of the quantum theory.

Soon, owing to Planck's laudatory comments and to experiments that gradually confirmed his theories, Einstein was invited to lecture at international meetings, such as the Solvay Conferences, and he rose rapidly in the academic world. He was offered a series of positions at increasingly prestigious institutions, including the University of Zürich, the University of Prague, the Swiss Federal Institute of Technology, and finally the University of Berlin, where he served as director of the Kaiser Wilhelm Institute for Physics from 1913 to 1933 (although the opening of the institute was delayed until 1917).

Even as his fame spread, Einstein's marriage was falling apart. He was constantly on the road, speaking at international conferences, and lost in contemplation of relativity. The couple argued frequently about their children and their meager finances. Convinced that his marriage was doomed, Einstein began an affair with a cousin, Elsa Löwenthal, whom he later married. (Elsa was a first cousin on his mother's side and a second cousin on his father's side.) When he finally divorced Mileva in 1919, he agreed to give her the money he might receive if he ever won a Nobel Prize.

One of the deep thoughts that consumed Einstein from 1905 to 1915 was a crucial flaw in his own theory: it made no mention of gravitation or acceleration. His friend Paul Ehrenfest had noticed a curious fact. If a disk is spinning, its rim travels faster than its centre, and hence (by special relativity) metre sticks placed on its circumference should shrink. This meant that Euclidean plane geometry must fail for the disk. For the next 10 years, Einstein would be absorbed with formulating a theory of gravity in terms of the curvature of space-time. To Einstein, Newton's gravitational force was actually a by-product of a deeper reality: the bending of the fabric of space and time.

In November 1915 Einstein finally completed the general theory of relativity, which he considered to be his masterpiece. In the summer of 1915, Einstein had given six two-hour lectures at the University of Göttingen that thoroughly explained general relativity, albeit with a few unfinished mathematical details. Much to Einstein's consternation, the mathematician David Hilbert, who had organized the lectures at his university, then completed these details and submitted a paper in November on general relativity just five days before Einstein, as if the theory were his own. Later they patched up their differences and remained friends. Einstein would write to Hilbert,

> *I struggled against a resulting sense of bitterness, and I did so with complete success. I once more think of you in unclouded friendship, and would ask you to try to do likewise toward me.*

Today physicists refer to the equations as the Einstein-Hilbert action, but the theory itself is attributed solely to Einstein.

Einstein was convinced that general relativity was correct because of its mathematical beauty and because it accurately predicted the perihelion of Mercury's orbit around the Sun. His theory also predicted a measurable deflection of light around the Sun. As a consequence, he even offered to help fund an expedition to measure the deflection of starlight during an eclipse of the Sun.

Einstein's work was interrupted by World War I. A lifelong pacifist, he was only one of four intellectuals in Germany to sign a manifesto opposing Germany's entry into war. Disgusted, he called nationalism "the measles of mankind." He would write, "At such a time as this, one realizes what a sorry species of animal one belongs to."

In the chaos unleashed after the war, in November 1918, radical students seized control of the University of Berlin and held the rector of the college and several professors hostage. Many feared that calling in the police to release the officials would result in a tragic confrontation. Einstein, because he was respected by both students and faculty, was the logical candidate to mediate this crisis. Together with Max Born, Einstein brokered a compromise that resolved it.

After the war, two expeditions were sent to test Einstein's prediction of deflected starlight near the Sun. One set sail for the island of Principe, off the coast of West Africa, and the other to Sobral in northern Brazil in order to observe the solar eclipse of May 29, 1919. On Nov. 6, 1919, the results were announced in London at a joint meeting of the Royal Society and the Royal Astronomical Society.

Nobel laureate J.J. Thomson, president of the Royal Society, stated:

> *This result is not an isolated one, it is a whole continent of scientific ideas....This is the most important result obtained in connection with the theory of gravitation since Newton's day, and it is fitting that it should be announced at a meeting of the Society so closely connected with him.*

The headline of *The Times* of London read, "Revolution in Science—New Theory of the Universe—Newton's Ideas Overthrown—Momentous Pronouncement—Space 'Warped.'" Almost immediately, Einstein became a world-renowned physicist, the successor to Isaac Newton.

Invitations came pouring in for him to speak around the world. In 1921 Einstein began the first of several world tours, visiting the United States, England, Japan, and France. Everywhere he went, the crowds numbered in the thousands. En route from Japan, he received word that he had received the Nobel Prize for Physics, but for the photoelectric effect rather than for his relativity theories. During his acceptance speech, Einstein startled the audience by speaking about relativity instead of the photoelectric effect.

Einstein also launched the new science of cosmology. His equations predicted that the universe is dynamic—expanding or contracting. This contradicted the prevailing view that the universe was static, so he reluctantly introduced a "cosmological term" to stabilize his model of the universe. In 1929 astronomer Edwin Hubble found that the universe was indeed expanding, thereby confirming Einstein's earlier work. In 1930, in a visit to the Mount Wilson Observatory near Los Angeles, Einstein met with Hubble and declared the cosmological constant to be his "greatest blunder." Recent satellite data, however, have shown that the cosmological constant is probably not zero but actually dominates the matter-energy content of the entire universe. Einstein's "blunder" apparently determines the ultimate fate of the universe.

During that same visit to California, Einstein was asked to appear alongside the comic actor Charlie Chaplin during the Hollywood debut of the film *City Lights*. When they were mobbed by thousands, Chaplin remarked, "The people applaud me because everybody understands me, and they applaud you because no one understands you." Einstein asked Chaplin, "What does it all mean?" Chaplin replied, "Nothing."

Einstein also began correspondences with other influential thinkers during this period. He corresponded with Sigmund Freud (both of them had sons with mental problems) on whether war was intrinsic to humanity. He discussed with the Indian mystic Rabindranath Tagore the question of whether consciousness can affect existence. One journalist remarked,

> *It was interesting to see them together—Tagore, the poet with the head of a thinker, and Einstein, the thinker with the head of a poet. It seemed to an observer as though two planets were engaged in a chat.*

Einstein also clarified his religious views, stating that he believed there was an "old one" who was the ultimate lawgiver. He wrote that he did not believe in a personal God that intervened in human affairs but instead believed in the God of the 17th-century Dutch Jewish philosopher Benedict de Spinoza—the God of harmony and beauty. His task, he believed, was to formulate a master theory that would allow him to "read the mind of God." He would write,

> *I'm not an atheist and I don't think I can call myself a pantheist. We are in the position of a little child entering a huge library filled with books in many different languages....The child dimly suspects a mysterious order in the arrangement of the books but doesn't know what it is. That, it seems to me, is the attitude of even the most intelligent human being toward God.*

Inevitably, Einstein's fame and the great success of his theories created a backlash. The rising Nazi movement found a convenient target in relativity, branding it "Jewish physics" and sponsoring conferences and book burnings to denounce Einstein and his theories. The Nazis enlisted other physicists, including Nobel laureates Philipp Lenard and Johannes Stark, to denounce Einstein. *One Hundred Authors Against Einstein* was published in 1931. When asked to comment on this denunciation of relativity by so many scientists, Einstein replied that to defeat relativity one did not need the word of 100 scientists, just one fact.

In December 1932 Einstein decided to leave Germany forever (he would never go back). It became obvious to Einstein that his life was in danger. A Nazi organization published a magazine with Einstein's picture and the caption "Not Yet Hanged" on the cover. There was even a price on his head. So great was the threat that Einstein split with his pacifist friends and said that it was justified to defend yourself with arms against Nazi aggression. To Einstein, pacifism was not an absolute concept but one that had to be re-examined depending on the magnitude of the threat.

Einstein settled at the newly formed Institute for Advanced Study at Princeton, N.J., which soon became a mecca for physicists from around the world. Newspaper articles declared that the "pope of physics" had left Germany and that Princeton had become the new Vatican.

The 1930s were hard years for Einstein. His son Eduard was diagnosed with schizophrenia and suffered a mental breakdown in 1930. (Eduard would be institutionalized for the rest of his life.) Einstein's close friend, physicist Paul Ehrenfest, who helped in the development of general relativity, committed suicide in 1933. And Einstein's beloved wife, Elsa, died in 1936.

To his horror, during the late 1930s, physicists began seriously to consider whether his equation $E = mc^2$ might make an atomic bomb possible. In 1920 Einstein himself had considered but eventually dismissed the possibility. However, he left it open if a method could be found to magnify the power of the atom. Then in 1938–39 Otto Hahn, Fritz Strassmann, Lise Meitner, and Otto Frisch showed that vast amounts of energy could be unleashed by the splitting of the uranium atom. The news electrified the physics community.

In July 1939 physicist Leo Szilard asked Einstein if he would write a letter to U.S. President Franklin D. Roosevelt urging him to develop an atomic bomb. Following several translated drafts, Einstein signed a letter on August 2 that was delivered to Roosevelt by one of his economic advisers, Alexander Sachs, on October 11. Roosevelt wrote back on October 19, informing Einstein that he had organized the Uranium Committee to study the issue.

Einstein was granted permanent residency in the United States in 1935 and became an American citizen in 1940, although he chose to retain his Swiss citizenship. During the war, Einstein's colleagues were asked to journey to the desert town of Los Alamos, N.M., to develop the first atomic bomb for the Manhattan Project. Einstein, the man whose equation had set the whole effort into motion, was never asked to participate. Voluminous declassified Federal Bureau of Investigation (FBI) files, numbering several thousand, reveal the reason: the U.S. government feared Einstein's lifelong association with peace and socialist organizations. (FBI director J. Edgar Hoover went so far as to recommend that Einstein be kept out of America by the Alien Exclusion Act, but he was overruled by the U.S. State Department.) Instead, during the war Einstein was asked to help the U.S. Navy evaluate designs for future weapons systems. Einstein also helped the war effort by auctioning off priceless personal manuscripts. In particular, a handwritten copy of his 1905

On his 70th birthday, Albert Einstein greeting children from the Reception Shelter of United Service for New Americans in New York City at his home in Princeton, N.J. Encyclopædia Britannica, Inc.

paper on special relativity was sold for $6.5 million. It is now located in the Library of Congress.

Einstein was on vacation when he heard the news that an atomic bomb had been dropped on Japan. Almost immediately he was part of an international effort to try to bring the atomic bomb under control, forming the Emergency Committee of Atomic Scientists.

The physics community split on the question of whether to build a hydrogen bomb. J. Robert Oppenheimer, the director of the atomic bomb project, was stripped of his security clearance for having suspected leftist associations. Einstein backed Oppenheimer and opposed the development of the hydrogen bomb, instead calling for international controls on the spread of nuclear technology. Einstein also was increasingly drawn to antiwar activities and to advancing the civil rights of African Americans.

In 1952 David Ben-Gurion, Israeli's premier, offered Einstein the post of president of Israel. Einstein, a prominent figure in the Zionist movement, respectfully declined.

Although Einstein continued to pioneer many key developments in the theory of general relativity—such as wormholes, higher dimensions, the possibility of time travel, the existence of black holes, and the creation of the universe—he was increasingly isolated from the rest of the physics community. Because of the huge strides made by quantum theory in unraveling the secrets of atoms and molecules, the majority of physicists were working on the quantum theory, not relativity. In fact, Einstein would engage in a series of historic private debates with Niels Bohr, originator of the Bohr atomic model. Through a series of sophisticated "thought experiments," Einstein tried to find logical inconsistencies in the quantum theory, particularly its lack of a deterministic mechanism. Einstein would often say that "God does not play dice with the universe."

In 1935 Einstein's most celebrated attack on the quantum theory led to the EPR (Einstein-Podolsky-Rosen) thought experiment. According to quantum theory, under certain circumstances two electrons separated by huge distances would have their properties linked, as if by an umbilical cord. Under these circumstances, if the properties of the first electron were measured, the state of the second electron would be known instantly—faster than the speed of light. This conclusion, Einstein claimed, clearly violated relativity. (Experiments conducted since then have confirmed that the quantum theory, rather than Einstein, was correct about the EPR experiment. In essence, what Einstein had actually shown was that quantum mechanics is nonlocal; i.e., random information can travel faster than light. This does not violate relativity, because the information is random and therefore useless.)

The other reason for Einstein's increasing detachment from his colleagues was his obsession, beginning in 1925, with discovering a unified field theory—an all-embracing theory that would unify the forces of the universe, and thereby the laws of physics, into one framework. In his later years he stopped opposing the quantum theory and tried to incorporate it, along with light and gravity, into a larger unified field theory. Gradually Einstein became set in his ways. He rarely traveled far and confined himself to long walks around Princeton with close associates, whom he engaged in deep conversations about politics, religion, physics, and his unified field theory. In 1950 he published an article on his theory in *Scientific American*, but because it neglected the still-mysterious strong force, it was necessarily incomplete. When he died five years later of an aortic aneurysm, it was still unfinished.

In some sense, Einstein, instead of being a relic, may have been too far ahead of his time. The strong force, a major piece of any unified field theory, was still a total mystery in Einstein's lifetime. Only in the 1970s and '80s did physicists begin to unravel the secret of the strong force with the quark model. Nevertheless, Einstein's work continues to win Nobel Prizes for succeeding physicists. In 1993 a Noble Prize was awarded to the discoverers of gravitation waves, predicted by Einstein. In 1995 a Nobel Prize was awarded to the discoverers of Bose-Einstein condensates (a new form of matter that can occur at extremely

Italian-born physicist Enrico Fermi explaining a problem in physics, c. 1950. National Archives, Washington, D.C.

low temperatures). Known black holes now number in the thousands. New generations of space satellites have continued to verify the cosmology of Einstein. And many leading physicists are trying to finish Einstein's ultimate dream of a "theory of everything."

ENRICO FERMI

(b. Sept. 29, 1901, Rome, Italy—d. Nov. 28, 1954, Chicago, Ill., U.S.)

Italian-born American scientist Enrico Fermi was one of the chief architects of the nuclear age. He developed the mathematical statistics required to clarify a large class of subatomic phenomena, explored nuclear transformations caused by neutrons, and directed the first controlled chain reaction involving nuclear fission. He was awarded the 1938 Nobel Prize for Physics, and the Enrico Fermi Award of the U.S. Department of Energy is given in his honour. Fermilab, the National Accelerator Laboratory in Illinois, is named for him, as is fermium, element number 100.

Fermi's father, Alberto Fermi, was a chief inspector of the government railways; his mother was Ida de Gattis, a schoolteacher. In 1918 Enrico Fermi won a scholarship to the University of Pisa's distinguished Scuola Normale Superiore, where his knowledge of recent physics benefited even the professors. After receiving a doctorate in 1922, Fermi used fellowships from the Italian Ministry of Public Instruction and the Rockefeller Foundation to study in Germany under Max Born at the University of Göttingen and in the Netherlands under Paul Ehrenfest at the State University of Leiden.

Fermi returned home to Italy in 1924 to a position as a lecturer in mathematical physics at the University of Florence. His early research was in general relativity, statistical mechanics, and quantum mechanics. Examples of gas degeneracy (appearance of unexpected phenomena) had been known, and some cases were explained by Bose-Einstein statistics, which describes the behaviour of subatomic particles known as bosons. Between 1926 and 1927, Fermi and the English physicist P.A.M. Dirac independently developed new statistics, now known as Fermi-Dirac statistics, to handle the subatomic particles that obey the Pauli exclusion principle; these particles, which include electrons, protons, neutrons (not yet discovered), and other particles with half-integer spin, are now known as fermions. This was a contribution of exceptional importance to atomic and nuclear physics, particularly in this period when quantum mechanics was first being applied.

This seminal work brought Fermi an invitation in 1926 to become a full professor at the University of Rome. Shortly after Fermi took up his new position in 1927, Franco Rasetti, a friend from Pisa and another superb experimentalist, joined Fermi in Rome, and they began to gather a group of talented students about them. These included Emilio Segrè, Ettore Majorana, Edoardo Amaldi, and Bruno Pontecorvo, all of whom had distinguished careers. Fermi, a charismatic, energetic, and seemingly infallible figure, clearly was the leader—so much so that his colleagues called him "the Pope."

In 1929 Fermi, as Italy's first professor of theoretical physics and a rising star in European science, was named by Italian Prime Minister Benito Mussolini to his new Accademia d'Italia, a position that included a substantial salary (much larger than that for any ordinary university position), a uniform, and a title ("Excellency").

During the late 1920s, quantum mechanics solved problem after problem in atomic physics. Fermi, earlier than most others, recognized that the field was becoming exhausted, however, and he deliberately changed his focus to the more primitively developed field of nuclear physics. Radioactivity had been recognized as a nuclear phenomenon for almost two decades by this time, but puzzles still abounded. In beta decay, or the expulsion of a negative electron from the nucleus, energy and momentum seemed not to be conserved. Fermi made use of the neutrino, an almost undetectable particle that had been postulated a few years earlier by the Austrian-born physicist Wolfgang Pauli, to fashion a theory of beta decay in which balance was restored. This led to recognition that beta decay was a manifestation of the weak force, one of the four known universal forces (the others being gravitation, electromagnetism, and the strong force).

In 1933 the French husband-and-wife team of Frédéric and Irène Joliot-Curie discovered artificial radioactivity caused by alpha particles (helium nuclei). Fermi quickly reasoned that the neutral neutron, found a year earlier by the English physicist James Chadwick, would be an even better projectile with which to bombard charged nuclei in order to initiate such reactions. With his colleagues, Fermi subjected more than 60 elements to neutron bombardment, using a Geiger-Müller counter to detect emissions and conducting chemical analyses to determine the new radioactive isotopes produced. Along the way, they found by chance that neutrons that had been slowed in their velocity often were more effective. When testing uranium they observed several activities, but they could not interpret what occurred. Some scientists thought that they had produced transuranium elements, namely elements higher than uranium at atomic number 92. The issue was not resolved until 1938, when the German chemists Otto Hahn and Fritz Strassmann experimentally, and the Austrian physicists Lise Meitner and Otto Frisch theoretically, cleared the confusion by revealing that the uranium had split and the several radioactivities detected were from fission fragments.

Fermi was little interested in politics, yet he grew increasingly uncomfortable with the fascist politics of his homeland. When Italy adopted the anti-Semitic policies of its ally, Nazi Germany, a crisis occurred, for Fermi's wife, Laura, was Jewish. The award of the 1938 Nobel Prize for Physics serendipitously provided the excuse for the family to travel abroad, and the prize money helped to establish them in the United States.

Settling first in New York City and then in Leonia, N.J., Fermi began his new life at Columbia University, in New York City. Within weeks of his arrival, news that uranium could fission astounded the physics community. Scientists had known for many years that nuclei could disgorge small chunks, such as alpha particles, beta particles, protons, and neutrons, either in natural radioactivity or upon bombardment by a projectile. However, they had never seen a nucleus split almost in two. The implications were both exciting and ominous, and they were recognized widely. When uranium fissioned, some mass was converted to energy, according to Albert Einstein's famous formula $E = mc^2$. Uranium also emitted a few neutrons in addition to the larger fragments. If these neutrons could be slowed to maximize their efficiency, they could participate in a controlled chain reaction to produce energy; that is, a nuclear reactor could be built. The same neutrons traveling at their initial high speed could also participate in an uncontrolled chain reaction, liberating an enormous amount of energy through many generations of fission events, all within a fraction of a second; that is, an atomic bomb could be built.

Working primarily with the Hungarian-born physicist Leo Szilard, Fermi constructed experimental arrangements of neutron sources and pieces of uranium. They sought to determine the necessary size of a structure, the best material to use as a moderator to slow neutrons, the necessary purity of all components (so neutrons would not be lost), and the best substance for forming control rods that could absorb neutrons to slow or stop the reaction. Fermi visited Washington, D.C., to alert the U.S. Navy about their research, but his guarded enthusiasm led only to a tiny grant. It was left to Einstein's letter to U.S. Pres. Franklin D. Roosevelt about the potential of an atomic bomb, in the summer of 1939, to initiate continuing government interest, and even that grew slowly.

When the United States entered World War II in December 1941, nuclear research was consolidated to some degree. Fermi had built a series of "piles," as he called them, at Columbia. Now he moved to the University of Chicago, where he continued to construct piles in a space under the stands of the football field. The final structure, a flattened sphere about 7.5 metres (25 feet) in diameter, contained 380 tons of graphite blocks as the moderator and 6 tons of uranium metal and 40 tons of uranium oxide as the fuel, distributed in a careful pattern. The pile went "critical" on Dec. 2, 1942, proving that a nuclear reaction could be initiated, controlled, and stopped. Chicago Pile-1, as it was called, was the first prototype for several large nuclear reactors constructed at Hanford, Wash., where plutonium, a man-made element heavier than uranium, was produced. Plutonium also could fission and thus was another route to the atomic bomb.

In 1944 Fermi became an American citizen and moved to Los Alamos, N.M., where physicist J. Robert Oppenheimer led the Manhattan Project's laboratory, whose mission was to fashion weapons out of the rare uranium-235 isotope and plutonium. Fermi was an associate director of the lab and headed one of its divisions. When the first plutonium bomb was tested on July 16, 1945, near Alamogordo, N.M., Fermi ingeniously made a rough calculation of its explosive energy by noting how far slips of paper were blown from the vertical.

After the war ended, Fermi accepted a permanent position at the University of Chicago, where he influenced another distinguished group of physicists, including Harold Agnew, Owen Chamberlin, Geoffrey Chew, James Cronin, Jerome Friedman, Richard Garwin, Murray Gell-Mann, Marvin Goldberger, Tsung-Dao Lee, Jack Steinberger, and Chen Ning Yang. As in Rome, Fermi recognized that his current pursuits, now in nuclear physics, were approaching a condition of maturity. He thus redirected his sights on reactions at higher energies, a field called elementary particle physics, or high-energy physics.

Since the war, science had been recognized in the United States as highly important to national security. Fermi largely avoided politics, but he did agree to serve on the General Advisory Committee (GAC), which counseled the five commissioners of the Atomic Energy Commission. In response to the revelation in September 1949 that the Soviet Union had detonated an atomic bomb, many Americans urged the government to try to construct a thermonuclear bomb, which can be orders of magnitude more powerful. GAC was publicly unanimous in opposing this step, mostly on technical grounds, with Fermi and Isidor Rabi going further by introducing an ethical question into so-called "objective" advice. Such a bomb, they wrote, "becomes a weapon which in practical effect is almost one of genocide.... It is necessarily an evil thing considered in any light." U.S. Pres. Harry S. Truman decided otherwise, and a loyal Fermi went for a time back to Los Alamos to assist in the development of fusion weapons, however with the hope that they might prove impossible to construct.

Fermi primarily investigated subatomic particles, particularly pi mesons and muons, after returning to Chicago. He was also known as a superb teacher, and many of his lectures are still in print. During his later years he raised a question now known as the Fermi paradox: "Where is everybody?" He was asking why no extraterrestrial civilizations seemed to be around to be detected, despite the great size and age of the universe. He pessimistically thought that the answer might involve nuclear annihilation.

RICHARD P. FEYNMAN

(b. May 11, 1918, New York, N.Y., U.S. — d. Feb. 15, 1988, Los Angeles, Calif.)

American theoretical physicist Richard Phillips Feynman was widely regarded as the most brilliant, influential, and iconoclastic figure in his field in the post-World War II era.

Feynman remade quantum electrodynamics—the theory of the interaction between light and matter—and thus altered the way science understands the nature of waves and particles. He was co-awarded the Nobel Prize for Physics in 1965 for this work, which tied together in an experimentally perfect package all the varied phenomena at work in light, radio, electricity, and magnetism. The other cowinners of the Nobel Prize, Julian S. Schwinger of the United States and Tomonaga Shin'ichirō of Japan, had independently created equivalent theories, but it was Feynman's that proved the most original and far-reaching. The problem-solving tools that he invented—including pictorial representations of particle interactions known as Feynman diagrams—permeated many areas of theoretical physics in the second half of the 20th century.

Born in the Far Rockaway section of New York City, Feynman was the descendant of Russian and Polish Jews who had immigrated to the United States late in the 19th century. He studied physics at the Massachusetts Institute of Technology, where his undergraduate thesis (1939) proposed an original and enduring approach to calculating forces in molecules. Feynman received his doctorate at Princeton University in 1942. At Princeton, with his adviser, John Archibald Wheeler, he developed an approach to quantum mechanics governed by the principle of least action. This approach replaced the wave-oriented electromagnetic picture developed by James Clerk Maxwell with one based entirely on particle interactions mapped in space and time. In effect, Feynman's method calculated the probabilities of all the possible paths a particle could take in going from one point to another.

During World War II Feynman was recruited to serve as a staff member of the U.S. atomic bomb project at Princeton University (1941–42) and then at the new secret laboratory at Los Alamos, N.M. (1943–45). At Los Alamos he became the youngest group leader in the theoretical division of the Manhattan Project. With the head of that division, Hans Bethe, he devised the formula for predicting the energy yield of a nuclear explosive. Feynman also took charge of the project's primitive computing effort, using a hybrid of new calculating machines and human workers to try to process the vast amounts of numerical computation required by the project. He observed the first detonation of an atomic bomb on July 16, 1945, near Alamogordo, and, though his initial reaction was euphoric, he later felt anxiety about the force he and his colleagues had helped unleash on the world.

At war's end Feynman became an associate professor at Cornell University (1945–50) and returned to studying the fundamental issues of quantum electrodynamics. In the years that followed, his vision of particle interaction kept returning to the forefront of physics as scientists explored esoteric new domains at the subatomic level. In 1950 he became professor of theoretical physics at the California Institute of Technology (Caltech), where he remained the rest of his career.

Five particular achievements of Feynman stand out as crucial to the development of modern physics. First, and most important, is his work in correcting the inaccuracies of earlier formulations of quantum electrodynamics, the theory that explains the interactions between electromagnetic radiation (photons) and charged subatomic particles such as electrons and positrons (antielectrons). By 1948 Feynman completed this reconstruction of a large part of quantum mechanics and electrodynamics and resolved the meaningless results that the old quantum electrodynamic theory sometimes produced. Second, he introduced simple diagrams, now called Feynman diagrams, that are easily visualized graphic analogues of the complicated mathematical expressions needed to describe the behaviour of systems of interacting particles. This work greatly simplified some of the calculations used to observe and predict such interactions.

In the early 1950s Feynman provided a quantum-mechanical explanation for the Soviet physicist Lev D. Landau's theory of superfluidity—i.e., the strange, frictionless behaviour of liquid helium at temperatures near absolute zero. In 1958 he and the American physicist Murray Gell-Mann devised a theory that accounted for most of the phenomena associated with the weak force, which is the force at work in radioactive decay. Their theory, which turns on the asymmetrical "handedness" of particle spin, proved particularly fruitful in modern particle physics. And finally, in 1968, while working with experimenters at the Stanford Linear Accelerator on the scattering of high-energy electrons by protons, Feynman invented a theory of "partons," or hypothetical hard particles inside the nucleus of the atom, that helped lead to the modern understanding of quarks.

Feynman's stature among physicists transcended the sum of even his sizable contributions to the field. His bold and colourful personality, unencumbered by false dignity or notions of excessive self-importance, seemed to announce: "Here is an unconventional mind." He was a master calculator who could create a dramatic impression in a group of scientists by slashing through a difficult numerical problem. His purely intellectual reputation became a part of the scenery of modern science. Feynman diagrams, Feynman integrals, and Feynman rules joined Feynman stories in the everyday conversation of physicists. They would say of a promising young colleague, "He's no Feynman, but..." His fellow physicists envied his flashes of inspiration and admired him for other qualities as well: a faith in nature's simple truths, a skepticism about official wisdom, and an impatience with mediocrity.

Feynman's lectures at Caltech evolved into the books *Quantum Electrodynamics* (1961) and *The Theory of Fundamental Processes* (1961). In 1961 he began reorganizing and teaching the introductory physics course at Caltech; the result, published as *The Feynman Lectures on Physics*, 3 vol. (1963–65), became a classic textbook. Feynman's views on quantum mechanics, scientific method, the relations between science and religion, and the role of beauty and uncertainty in scientific knowledge are expressed in two models of science writing, again distilled from lectures: *The Character of Physical Law* (1965) and *QED: The Strange Theory of Light and Matter* (1985).

ALEKSANDR ALEKSANDROVICH FRIEDMANN

(b. June 17 [June 29, New Style], 1888, St. Petersburg, Russia—d. Sept. 16, 1925, Leningrad [St. Petersburg])

Aleksandr Aleksandrovich Friedmann was a Russian mathematician and physical scientist.

After graduating from the University of St. Petersburg in 1910, Friedmann joined the Pavlovsk Aerological Observatory and, during World War I, did aerological work for the Russian army. After the war he was on the staff of the University of Perm (1918–20) and then on the staffs of the Main Physical Observatory and other institutions until his death in 1925.

In 1922–24 Friedmann used Einstein's general theory of relativity to formulate the mathematics of a dynamic (time-dependent) universe. (Einstein and Dutch mathematician Willem de Sitter had earlier studied static cosmologies.) In the Friedmann models, the average mass density is constant over all space but may change with time as the universe expands. His models, which included all three cases of positive, negative, and zero curvature, were crucial in the development of modern cosmology. Friedmann also calculated the time back to the moment when an expanding universe would have been a mere point, obtaining tens of billions of years; but it is not clear how much physical significance he attributed to this speculation. It may, however, still be considered a part of the prehistory of the big-bang theory. Friedmann also considered the possibility of a cyclical universe. In his other work, he was among the founders of the science of dynamic meteorology.

GEORGE GAMOW

(b. March 4, 1904, Odessa, Russian Empire [now in Ukraine]—d. Aug. 19, 1968, Boulder, Colo., U.S.)

Russian-born American nuclear physicist and cosmologist George Gamow was one of the foremost advocates of the big-bang theory, according to which the universe was formed in a colossal explosion that took place billions of years ago. In addition, his work on deoxyribonucleic acid (DNA) made a basic contribution to modern genetic theory.

Gamow attended Leningrad (now St. Petersburg) University, where he studied briefly with A.A. Friedmann, a mathematician and cosmologist who suggested that the universe should be expanding. At that time Gamow did not pursue Friedmann's suggestion, preferring instead to delve into quantum theory. After graduating in 1928, he traveled to Göttingen, where he developed his quantum theory of radioactivity, the first successful explanation of the behaviour of radioactive elements, some of which decay in seconds while others decay over thousands of years.

His achievement earned him a fellowship at the Copenhagen Institute of Theoretical Physics (1928–29), where he continued his investigations in theoretical nuclear physics. There he proposed his "liquid drop" model of atomic nuclei, which served as the basis for the modern theories of nuclear fission and fusion. He also collaborated with F. Houtermans and R. Atkinson in developing a theory of the rates of thermonuclear reactions inside stars.

In 1934, after emigrating from the Soviet Union, Gamow was appointed professor of physics at George Washington University in Washington, D.C. There he collaborated with Edward Teller in developing a theory of beta decay (1936), a nuclear decay process in which an electron is emitted.

Soon after, Gamow resumed his study of the relations between small-scale nuclear processes and cosmology. He used his knowledge of nuclear reactions to interpret stellar evolution, collaborating with Teller on a theory of the internal structures of red giant stars (1942). From his work on stellar evolution, Gamow postulated that the Sun's energy results from thermonuclear processes.

Gamow and Teller were both proponents of the expanding-universe theory that had been advanced by Friedmann, Edwin Hubble, and Georges LeMaître. Gamow, however, modified the theory, and he, Ralph Alpher, and Hans Bethe published this theory in a paper called "The Origin of Chemical Elements" (1948). This paper, attempting to explain the distribution of chemical elements throughout the universe, posits a primeval thermonuclear explosion, the big bang that began the universe. According to the theory, after the big bang, atomic nuclei were built up by the successive capture of neutrons by the initially formed pairs and triplets.

In 1954 Gamow's scientific interests grew to encompass biochemistry. He proposed the concept of a genetic code and maintained that the code was determined by the order of recurring triplets of nucleotides, the basic components of DNA. His proposal was vindicated during the rapid development of genetic theory that followed.

Gamow held the position of professor of physics at the University of Colorado, Boulder, from 1956 until his death. He is perhaps best known for his popular writings, designed to introduce to the nonspecialist such difficult subjects as relativity and cosmology. His first such work, *Mr. Tomkins in Wonderland* (1936), gave rise to the multivolume "Mr. Tomkins" series (1939–67). Among his other writings are *One, Two, Three...Infinity* (1947), *The Creation of the Universe* (1952; rev. ed., 1961), *A Planet Called Earth* (1963), and *A Star Called the Sun* (1964).

HANS GEIGER

(b. Sept. 30, 1882, Neustadt an der Haardt, Ger.—d. Sept. 24, 1945, Potsdam)

German physicist Hans Geiger introduced the first successful detector (the Geiger counter) of individual alpha particles and other ionizing radiations.

Geiger was awarded Ph.D. by the University of Erlangen in 1906 and shortly thereafter joined the staff of the University of Manchester, where he became one of the most valuable collaborators of Ernest Rutherford. At Manchester, Geiger built the first version of his particle counter and used it and other radiation detectors in experiments that led to the identification of the alpha particle as the nucleus of the helium atom and to Rutherford's correct proposal (1912) that, in any atom, the nucleus occupies a very small volume at the centre.

Moving in 1912 to the Physikalisch-Technische Reichsanstalt ("German National Institute for Science and Technology") in Berlin, Geiger continued his studies of atomic structure. During World War I he served as an artillery officer in the German army. With Walther Bothe, Geiger devised the technique of coincidence counting and used it in 1924 to clarify the details of the Compton effect. In 1925 Geiger accepted his first teaching position, at the University of Kiel. There, he and Walther Müller improved the sensitivity, performance, and durability of the particle counter; the Geiger-Müller counter detects not only alpha particles but beta particles (electrons) and ionizing electromagnetic photons. In 1929 Geiger took up a post at the University of Tübingen, where he made his first observation of a cosmic-ray shower. He continued to investigate cosmic rays, artificial radioactivity, and nuclear fission after accepting a position in 1936 at the Technische Hochschule in Berlin, which he held until he died.

MURRAY GELL-MANN

(b. Sept. 15, 1929, New York, N.Y., U.S.)

American physicist Murray Gell-Mann won the Nobel Prize for Physics in 1969 for his work pertaining to the classification of subatomic particles and their interactions.

Having entered Yale University at age 15, Gell-Mann received his B.S. in physics in 1948 and his Ph.D. at the Massachusetts Institute of Technology in 1951. His doctoral research on subatomic particles was influential in the later work of the Nobel laureate (1963) Eugene P. Wigner. In 1952 Gell-Mann joined the Institute for Nuclear Studies at the University of Chicago. The following year he introduced the concept of "strangeness," a quantum property that accounted for previously puzzling decay patterns of certain mesons. As defined by Gell-Mann, strangeness is conserved when any subatomic particle interacts via the strong force—i.e., the force that binds the components of the atomic nucleus. Gell-Mann joined the faculty of the California Institute of Technology in Pasadena in 1955 and was appointed the Robert Andrews Millikan Professor of Theoretical Physics in 1967 (emeritus, 1993).

In 1961 Gell-Mann and Yuval Ne'eman, an Israeli theoretical physicist, independently proposed a scheme for classifying previously discovered strongly interacting particles into a simple, orderly arrangement of families. Called the Eightfold Way (after Buddha's Eightfold Path to Enlightenment and bliss), the scheme grouped mesons and baryons (e.g., protons and neutrons) into multiplets of 1, 8, 10, or 27 members on the basis of various properties. All particles in the same multiplet are to be thought of as variant states of the same basic particle. Gell-Mann speculated that it should be possible to explain certain properties of known particles in terms of even more fundamental particles, or building blocks. He later called these basic bits of matter "quarks," adopting the fanciful term from James Joyce's novel *Finnegans Wake*. One of the early successes of Gell-Mann's quark hypothesis was the prediction and subsequent discovery of the omega-minus particle (1964). Over the years, research has yielded other findings that have led to the wide acceptance and elaboration of the quark concept.

Gell-Mann published a number of works on this phase of his career, notable among which are *The Eightfold Way* (1964), written in collaboration with Ne'eman, and *Broken Scale Variance and the Light Cone* (1971), coauthored with K. Wilson.

In 1984 Gell-Mann cofounded the Santa Fe Institute, a nonprofit centre located in Santa Fe, N.M., that supports research concerning complex adaptive systems and emergent phenomena associated with complexity. In "Let's Call It Plectics," a 1995 article in the institute's journal, *Complexity*, he coined the word *plectics* to describe the type of research supported by the institute. In *The Quark and the Jaguar* (1994), Gell-Mann gave a fuller description of the ideas concerning the relationship between the basic laws of physics (the quark) and the emergent phenomena of life (the jaguar).

WALTHER GERLACH

(b. Aug. 1, 1889, Biebrich am Rhein, Ger.—d. Aug. 10, 1979, Munich)

German physicist Walther Gerlach was noted especially for his work with Otto Stern on the deflections of atoms in a nonhomogeneous magnetic field.

Educated at the University of Tübingen, he became a lecturer there in 1916; after periods at Göttingen and Frankfurt, he returned to Tübingen as professor of physics in 1925 and from 1929 to 1957 was professor of physics at Munich. He was best known for his part in the Stern-Gerlach experiment, but he also made contributions in the fields of radiation, spectroscopy, and quantum theory. His books include *Grundlagen der Quantentheorie* (1921), *Magnetismus* (1931), *Humaniora und Natur* (1950), and *Kepler und die Copernicanische Wende* (1973).

LESTER HALBERT GERMER

(b. Oct. 10, 1896, Chicago, Ill., U.S.—d. Oct. 3, 1971, Gardiner, N.Y.)

American physicist Lester Halbert Germer, with his colleague Clinton Joseph Davisson, conducted an experiment (1927) first demonstrated the wave properties of the electron. This experiment confirmed the hypothesis of Louis-Victor de Broglie, a founder of wave mechanics, that the electron should show the properties of an electromagnetic wave as well as those of a particle.

Germer was a graduate student at Columbia University, working under Davisson's supervision at the Bell Telephone Laboratories in New York City, when they bombarded a single crystal of nickel with an electron beam and observed that the distribution of the scattered electrons conformed closely to the prediction of de Broglie's hypothesis.

SAMUEL ABRAHAM GOUDSMIT

(b. July 11, 1902, The Hague, Neth.—d. Dec. 4, 1978, Reno, Nev., U.S.)

Dutch-born U.S. physicist Samuel Abraham Goudsmit, with George E. Uhlenbeck, a fellow graduate student at the University of Leiden, Neth., formulated (1925) the concept of electron spin, leading to major changes in atomic theory and quantum mechanics. Of this work Isidor I. Rabi, a Nobelist in physics, remarked, "Physics must be forever in debt to those two men for discovering the spin." Later it was recognized that spin is a fundamental property of neutrons, protons, and other elementary particles.

A faculty member of the University of Michigan (1927–46) and Northwestern University, Ill. (1946–48), Goudsmit worked on radar research at the Massachusetts Institute of Technology, Cambridge (1941–44), and was head of Alsos, a secret mission that followed the advancing Allied forces in Europe to determine the progress of Germany's atomic bomb project.

From 1948 to 1970 Goudsmit was a member of the staff of Brookhaven National Laboratory, Upton, N.Y., and then joined the University of Reno, Nevada. His works include *The Structure of Line Spectra*, with Linus Pauling (1930); *Atomic Energy States*, with Robert F. Bacher (1932); *Alsos* (1947); and *Time*, with Robert Claiborne (1966).

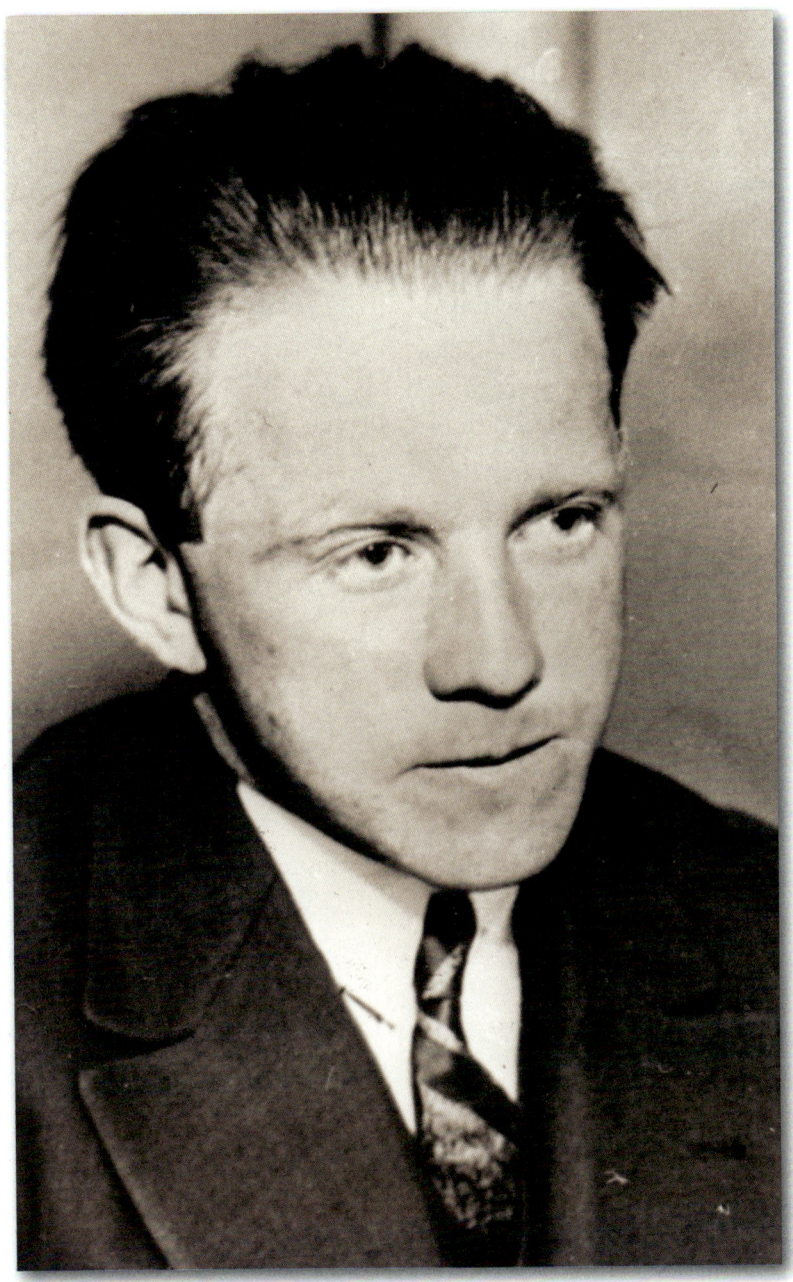

Werner Heisenberg developed the uncertainty principle of quantum mechanics. Keystone/Hulton Archive/Getty Images

WERNER HEISENBERG

(b. Dec. 5, 1901, Würzburg, Ger.—d. Feb. 1, 1976, Munich, W.Ger.)

German physicist and philosopher Werner Karl Heisenberg discovered (1925) a way to formulate quantum mechanics in terms of matrices. For that discovery, he was awarded the Nobel Prize for Physics in 1932. In 1927 he published his uncertainty principle, upon which he built his philosophy and for which he is best known. He also made important contributions to the theories of the hydrodynamics of turbulent flows, the atomic nucleus, ferromagnetism, cosmic rays, and subatomic particles, and he was instrumental in planning the first West German nuclear reactor at Karlsruhe, together with a research reactor in Munich, in 1957. Considerable controversy surrounds his work on atomic research during World War II.

Heisenberg's father, August Heisenberg, a scholar of ancient Greek philology and modern Greek literature, was a teacher at a gymnasium and lecturer at the University of Würzburg. Werner's mother, née Anna Wecklein, was the daughter of the rector of the elite Maximilians-Gymnasium in Munich. In 1910 August Heisenberg became a professor of Greek philology at the University of Munich. Werner entered the Maximilians-Gymnasium the following year and soon impressed his teachers with his precocity in mathematics. Heisenberg entered the University of Munich in 1920, becoming a student of Arnold Sommerfeld, an expert on atomic spectroscopy and exponent of the quantum model of physics. (The idea that certain properties in atomic physics are not continuous and take on only certain discrete, or quantized, values at small scales had been developed by Danish physicist Niels Bohr in 1913.) Heisenberg finished his formal work for a doctorate in 1923 with a dissertation on hydrodynamics.

Despite a mediocre dissertation defense, Heisenberg's real talents emerged in his work on the anomalous Zeeman effect, in which atomic spectral lines are split into multiple components under the influence of a magnetic field. Heisenberg developed a model that accounted for this phenomenon, though at the cost of introducing half-integer quantum numbers, a notion at odds with Bohr's theory as understood to date. While still officially Sommerfeld's student, in 1922 Heisenberg became an assistant and student of Max Born at the University of Göttingen, where Heisenberg also first met Bohr. In 1924 Heisenberg completed his habilitation, the qualification to teach at the university level in Germany.

In 1925, after an extended visit to Bohr's Institute of Theoretical Physics at the University of Copenhagen, Heisenberg tackled the problem of spectrum intensities of the electron taken as an anharmonic oscillator (a one-dimensional vibrating system). His position that the theory should be based only on observable quantities was central to his paper of July 1925, "Über quantentheoretische Umdeutung kinematischer und mechanischer Beziehungen" ("Quantum-Theoretical Reinterpretation of Kinematic and Mechanical Relations"). Heisenberg's formalism rested upon noncommutative multiplication; Born, together with his new assistant Pascual Jordan, realized that this could be expressed using matrix algebra, which they used in a paper submitted for publication in September as "Zur Quantenmechanik" ("On Quantum Mechanics"). By November, Born, Heisenberg, and Jordan had completed "Zur Quantenmechanik II" ("On Quantum Mechanics II"), colloquially known as the "three-man paper," which is regarded as the foundational document of a new quantum mechanics.

Other formulations of quantum mechanics were being devised during the 1920s: the bracket notation (using vectors in a Hilbert space) was developed by P.A.M. Dirac in England and the wave equation was worked out by Erwin Schrödinger in Switzerland (where the Austrian physicist was then working). Schrödinger soon demonstrated that the different formulations were mathematically equivalent, though the physical significance of this equivalence remained unclear. Heisenberg again returned to Bohr's institute in Copenhagen, and their conversations on this topic culminated in Heisenberg's landmark paper of March 1927, "Über den anschulichen Inhalt der quantentheoretischen Kinematik und Mechanik" ("On the Perceptual Content of Quantum Theoretical Kinematics and Mechanics").

This paper articulated the uncertainty, or indeterminacy, principle. Quantum mechanics demonstrated, according to Heisenberg, that the momentum (p) and position (q) of a particle could not both be exactly measured simultaneously. Instead, a relation exists between the indeterminacies (Δ) in the measurement of these variables such that $\Delta p \Delta q \geq h/4\pi$ (where h is Planck's constant, or about $6.62606896 \times 10^{-34}$ joule-second). Since there exists a lower limit ($h/4\pi$) on the product of the uncertainties, if the uncertainty in one variable diminishes toward 0, the uncertainty in the other must increase reciprocally. An analogous relation exists between any pair of canonically conjugate variables, such as energy and time.

Heisenberg drew a philosophically profound conclusion: absolute causal determinism was impossible, since it required exact knowledge of both position and momentum as initial conditions. Therefore, the use of probabilistic formulations in atomic theory resulted not from ignorance but from the necessarily indeterministic relationship between the variables. This viewpoint was central to the so-called "Copenhagen interpretation" of quantum theory, which got its name from the strong defense for the idea at Bohr's institute in Copenhagen. Although this became a predominant viewpoint, several leading physicists, including Schrödinger and Albert Einstein, saw the renunciation of deterministic causality as physically incomplete.

In 1927 Heisenberg took up a professorship in Leipzig. In exchange with Dirac, Jordan, Wolfgang Pauli, and others, he embarked on a research program to create a quantum field theory, uniting quantum mechanics with relativity theory to comprehend the interaction of particles and (force) fields. Heisenberg also worked on the theory of the atomic nucleus following the discovery of the neutron in 1932, developing a model of proton and neutron interaction through what came to be known as the strong force. The 1932 Nobel Prize for Physics was not announced until November 1933, when the 1933 winners were also announced. Heisenberg was awarded the 1932 physics prize, while Schrödinger and Dirac shared the 1933 physics prize.

The same year that Heisenberg was awarded a Nobel Prize, 1933, also saw the rise to power of the National Socialist German Workers' Party (Nazi Party). Nazi policies excluding "non-Aryans" or the politically "unreliable" from the civil service meant the dismissal or resignation of many professors and academics—including, for example, Born, Einstein, and Schrödinger and several of Heisenberg's students and colleagues in Leipzig. Heisenberg's response was mostly quiet interventions within the bureaucracy rather than overt public protest, guided by a hope that the Nazi regime or its most extreme manifestations would not last long.

Heisenberg also became the target of ideological attacks. A coterie of Nazi-affiliated physicists promoted the idea of a "German" or "Aryan" physics, opposed to a supposedly "Jewish" influence manifested in abstract mathematical approaches—above all, relativity and quantum theories. Johannes Stark, a leader of this movement, used his Nazi Party connections to assert influence over science funding and personnel decisions. Sommerfeld had long regarded Heisenberg as his eventual successor, and in 1937 Heisenberg received a call to join the University of Munich. Thereupon the official SS journal published an article signed by Stark that called Heisenberg a "white Jew" and the "Ossietzky of physics." (German journalist and pacifist Carl von Ossietzky, winner of the 1935 Nobel Prize for Peace, had been imprisoned in 1931 for treason for his reporting of Germany's secret rearmament efforts, given amnesty in 1932, and then rearrested and interned in a concentration camp by the Nazis in 1933.)

Heisenberg, relying on the coincidence that his mother's family was acquainted with Heinrich Himmler's family, sent a request to the SS chief to intervene in his behalf in acquiring the professorship in Munich. Himmler, after an investigation, decreed a compromise: Heisenberg would not succeed Sommerfeld in Munich, but he would be spared further personal attacks and (essentially) promised another prominent post in the future. Meanwhile, Stark and the Aryan physicists were for other reasons losing influence in the bureaucratic jungle of the Nazi state, particularly in the context of militarization. Amid this political turbulence, Heisenberg apparently never seriously contemplated leaving Germany, though he certainly received several offers of university appointments in the United States and elsewhere. Apparently, he was guided by a strong sense of personal duty to the profession and a national loyalty that (in his mind) transcended the particular politics of the regime.

In 1937 Heisenberg married Elisabeth Schumacher, the daughter of an economics professor, whom he had met at a concert. Twins were born the next year, the first of eventually seven children for the couple.

Heisenberg's main focus of work in the late 1930s was high-energy cosmic rays, for which he proposed a theory of "explosion showers," in which multiple particles were produced in a single process, in contrast to the "cascade" theory principally favoured by British and American physicists. Heisenberg also saw in cosmic ray phenomena possible evidence for his idea of a minimum length marking a lower boundary of the domain of quantum mechanics.

The discovery of nuclear fission pushed the atomic nucleus into the centre of attention. After the German invasion of Poland in 1939, Heisenberg was drafted to work for the Army Weapons Bureau on the problem of nuclear energy. At first commuting between Leipzig and the Kaiser Wilhelm Institute (KWI) for Physics in Berlin and, after 1942, as director at the latter, Heisenberg took on a leading role in Germany's nuclear research. Given the Nazi context, this role has been enormously controversial. Heisenberg's research group was unsuccessful, of course, in producing a reactor or an atomic bomb. In explanation, some accounts have presented Heisenberg as simply incompetent; others, conversely, have suggested that he deliberately delayed or sabotaged the effort. It is clear in retrospect that there were indeed critical mistakes at several points in the research. Likewise, it is apparent that the German nuclear weapons project as a whole was not possessed of the same degree of enthusiasm that pervaded the Manhattan Project in the United States. However, factors outside Heisenberg's direct control had a more substantive role in the outcome.

In contrast to the unified Anglo-American effort, the German project was bureaucratically fractured and cut off from international collaboration. Key materials were in short supply in Germany, to say nothing of the widespread dislocations caused by Allied bombing of the country's transportation network. Moreover, the overall strategic perspective critically affected the prioritization or de-prioritization of nuclear bomb research. After a 1942 conference with Axis scientists, German minister for armaments and war production Albert Speer concluded that reactor research should proceed but that any bomb was unlikely to be developed in time for use in the war. By way of confirmation, the official start of the Manhattan Project in the United States also occurred in 1942, and, even with its massive effort, it could not produce an atomic bomb before Germany's surrender.

Controversy has also swirled around Heisenberg's lectures in countries such as Denmark and the Netherlands during the war years. These trips outside of Germany were necessarily taken with the approval of German authorities and hence were perceived by colleagues in the occupied countries as indicating Nazi leaders' endorsement of Heisenberg and vice versa. Most notorious in this regard was a trip to Copenhagen in September 1941, during which Heisenberg raised the subject of nuclear weapons research in a conversation with Bohr, offending and alarming the latter, though Heisenberg later claimed that Bohr's reaction rested on some misunderstanding. The exact content of the conversation has never been clarified.

By January 1945 the KWI for Physics was evacuated to the towns of Hechingen and Haigerloch in the province of Hohenzollern (then a Prussian enclave, now part of the state of Baden-Württemberg). In the closing days of the war, Heisenberg bicycled from there to his family's vacation house in Bavaria. There he was captured by an American military intelligence team, and eventually he was interned with several other German physicists in England. Their conversations after news of the atomic bombing of Hiroshima, Japan, initially suggested that Heisenberg had no clear sense of some basic principles of bomb design—e.g., the approximate critical mass—but within a few days he had solved many of these problems.

Heisenberg was released by the British authorities in January 1946, and soon thereafter he resumed his directorship of the reconstituted Kaiser Wilhelm, which was soon renamed the Max Planck Institute for Physics, now in Göttingen. In the postwar years, Heisenberg took on a variety of roles as an administrator of and spokesman for German science within the Federal Republic of Germany, a shift to a more overtly political role that was in some contrast to his more apolitical stance before 1945. In 1949 Heisenberg became the first president of the German Research Council, a consortium of the Max Planck Society and the various West German academies of science that sought to promote German science in the international arena and to influence federal science funding through the newly elected chancellor Konrad Adenauer. However, this new organization encountered conflict with the older, now re-established Emergency Association for German Science, whose approach preserved the traditional primacy of the various German states in cultural and educational matters. In 1951 the Research Council merged with the Emergency Association to form the German Research Association. Beginning in 1952, Heisenberg was instrumental in Germany's participation in the creation of the European Council for Nuclear Research (CERN). In 1953 Heisenberg became the founding president of the third iteration of the Humboldt Foundation, a government-funded organization that provided fellowships for foreign scholars to conduct research in Germany. Despite these close connections with the federal government, Heisenberg also became an overt critic of Adenauer's policies as one of the "Göttingen 18" in 1957; following the government's announcement that it was considering equipping the army with (American-built) nuclear weapons, this group of nuclear scientists issued a manifesto protesting the plan.

In the postwar period Heisenberg continued his search for a comprehensive quantum field theory, utilizing the "scattering matrix" approach (first introduced in 1942) and returning to the notion of a minimum universal length as a key feature. In 1958 he proposed a unified field theory—newspaper stories referred to his "world formula"—which he saw as a symmetry-based approach to the proliferation of particles then under way. However, support from the physics community was limited, particularly with the appearance of the quark model in the 1960s. In 1958 Heisenberg also finally achieved the goal of an academic position in Munich, as the Max Planck Institute for Physics moved there in that year. Heisenberg retired from his institute directorship in 1970.

PASCUAL JORDAN

(b. Oct. 18, 1902, Hannover, Ger.—d. July 31, 1980, Hamburg)

German theoretical physicist Ernst Pascual Jordan was one of the founders of quantum mechanics and quantum field theory.

Jordan received a doctorate (1924) from the University of Göttingen, working with German physicists Max Born and James Franck on the problems of quantum theory. In 1925 Jordan published two seminal papers, one in collaboration with Born and German physicist Werner Heisenberg and one with just Born, that developed Heisenberg's initial idea of noncommutative variables into a formulation of quantum theory in terms of matrix mechanics—the first working version of quantum mechanics. In the following years, in Göttingen and as a Rockefeller fellow in Copenhagen, Jordan helped propel the new theory toward completion, incorporating the wave mechanics approach of the German physicist Erwin Schrödinger with the matrix formulation. The comprehensive mathematical formalism of nonrelativistic quantum mechanics was achieved for the first time in the transformation theory published by Jordan and independently by the English physicist P.A.M. Dirac in 1927.

Jordan also did pioneering work on the relativistic generalization of quantum mechanics and its application to electromagnetic radiation. In 1925 he used matrix mechanics to quantize electromagnetic waves. This method was further developed to great success in Dirac's 1927 paper on the quantum theory of radiation, in which also the idea of a second quantization (many-body formalism) for bosons made its first appearance. Jordan then put forward the general program of quantum field theory, proposing that relativistic quantum theory should describe all subatomic particles—matter and radiation alike—as quanta of wave fields. Working toward the implementation of this idea, he and the Hungarian-born American physicist Eugene P. Wigner showed in 1928 how the second quantization is capable of describing fermions, in addition to bosons, by introducing the technical idea of an anticommutator (a special matrix operator).

Heisenberg and the Austrian physicist Wolfgang Pauli completed the program in 1929–30, but their quantum electrodynamics theory almost immediately faced new difficulties and inspired a search for additional ideas. In the 1930s Jordan suggested further radicalizing mathematical formalism by using nonassociative variables (variables that do not obey the associative law). His proposal did not manage to help quantum field theory but did result in the development of (nonassociative) Jordan algebras in mathematics. In his later research, Jordan also worked on the application of quantum theory to biological problems, and he originated (concurrently with the American physicist Robert Dicke) a theory of cosmology that proposed to make the universal constants of nature variable and dependent upon the expansion of the universe.

Jordan was a professor of theoretical physics at the University of Rostock from 1928 to 1944. Although some of his closest professional friends and colleagues were Jewish, he joined the National Socialist German Worker's Party (Nazi Party) in 1933, when Adolf Hitler came to power. In his popular writings about science, Jordan argued that modern physics, including relativity and quantum mechanics, is ideologically compatible with National Socialism. During World War II he performed military research for the Luftwaffe (German air force). Jordan then became a professor at the Humboldt University of Berlin (1944–51) and the University of Hamburg (1951–71) in West Germany. He also served in the West German Bundestag (1957–61), representing the conservative Christian Democratic Union.

BRIAN D. JOSEPHSON

(b. Jan. 4, 1940, Cardiff, Glamorgan, Wales)

Brian David Josephson was a British physicist whose discovery of the Josephson effect while a 22-year-old graduate student won him a share (with Leo Esaki and Ivar Giaever) of the 1973 Nobel Prize for Physics.

He entered Trinity College, Cambridge, in pursuit of an education in physics and received his bachelor's (1960) and master's and Ph.D. degrees (1964) there, publishing his first work while still an undergraduate; it dealt with certain aspects of the special theory of relativity and the Mossbauer effect. He was elected a fellow of Trinity College in 1962. He was a brilliant and assured student; one former lecturer recalled a special need for precision in any presentation to a class that included Josephson—otherwise, the student would confront the instructor politely after class and explain the mistake.

While still an undergraduate, Josephson became interested in superconductivity, and he began to explore the properties of a junction between two superconductors that later came to be known as a Josephson junction. Josephson extended earlier work in tunneling, the phenomenon by which electrons functioning as radiated waves can penetrate solids, done by L. Esaki and I. Giaever. He showed theoretically that tunneling between two superconductors could have very special characteristics, e.g., flow across an insulating layer without the application of a voltage; if a voltage is applied, the current stops flowing and oscillates at high frequency. This was the Josephson effect. Experimentation confirmed it, and its confirmation in turn reinforced the earlier BCS theory of superconductor behaviour. Applying Josephson's discoveries with superconductors, researchers at International Business Machines Corporation had assembled by 1980 an experimental computer switch structure, which would permit switching speeds from 10 to 100 times faster than those possible with conventional silicon-based chips, increasing data processing capabilities by a vast amount.

He went to the United States to be a research professor at the University of Illinois in 1965–66 and in 1967 returned to Cambridge as assistant director of research.

He was appointed reader in physics in 1972 and professor of physics in 1974. He was elected a fellow of the Royal Society in 1970.

A few years before the Nobel award, Josephson grew interested in the possible relevance of Eastern mysticism to scientific understanding. In 1980 he and V.S. Ramachandran published an edited transcript of a 1978 international symposium on consciousness at Oxford under the title *Consciousness and the Physical World*.

MAX VON LAUE

(b. Oct. 9, 1879, Pfaffendorf, near Koblenz, Ger.—d. April 23, 1960, Berlin, W.Ger.)

German physicist Max Theodor Felix von Laue was the recipient of the Nobel Prize for Physics in 1914 for his discovery of the diffraction of X-rays in crystals. This enabled scientists to study the structure of crystals and hence marked the origin of solid-state physics, an important field in the development of modern electronics.

Laue became professor of physics at the University of Zürich in 1912. Laue was the first to suggest the use of a crystal to act as a grating for the diffraction of X-rays, showing that if a beam of X-rays passed through a crystal, diffraction would take place and a pattern would be formed on a photographic plate placed at a right angle to the direction of the rays. The pattern would mark out the symmetrical arrangements of the atoms in the crystal. This was verified experimentally in 1912 by two of Laue's students working under his direction. This success demonstrated that X-rays are electromagnetic radiations similar to light and also provided experimental proof that the atomic structure of crystals is a regularly repeating arrangement.

Laue championed Albert Einstein's theory of relativity, did research on the quantum theory, the Compton effect (change of wavelength in light under certain conditions), and the disintegration of atoms. He became director of the Institute for Theoretical Physics at the University of Berlin in 1919 and director of the Max Planck Institute for Research in Physical Chemistry, Berlin, in 1951.

HENDRIK ANTOON LORENTZ

(b. July 18, 1853, Arnhem, Neth.—d. Feb. 4, 1928, Haarlem)

Dutch physicist Hendrik Antoon Lorentz was the joint winner (with Pieter Zeeman) of the Nobel Prize for Physics in 1902 for his theory of electromagnetic radiation, which, confirmed by findings of Zeeman, gave rise to Albert Einstein's special theory of relativity.

In his doctoral thesis at the University of Leiden (1875), Lorentz refined the electromagnetic theory of James C. Maxwell of England so that it more satisfactorily explained the reflection and refraction of light. He was appointed professor of mathematical physics at Leiden in 1878. His work in physics was wide in scope, but his central aim was to construct a single theory to explain the relationship of electricity, magnetism, and light. Although, according to Maxwell's theory, electromagnetic radiation is produced by the oscillation of electric charges, the charges that produce light were unknown. Since it was generally believed that an electric current was made up of charged particles, Lorentz later theorized that the atoms of matter might also consist of charged particles and suggested that the oscillations of these charged particles (electrons) inside the atom were the source of light. If this were true, then a strong magnetic field ought to have an effect on the oscillations and therefore on the wavelength of the light thus produced. In 1896 Zeeman, a pupil of Lorentz, demonstrated this phenomenon, known as the Zeeman effect, and in 1902 they were awarded the Nobel Prize.

Lorentz' electron theory was not, however, successful in explaining the negative results of the Michelson-Morley experiment, an effort to measure the velocity of the Earth through the hypothetical luminiferous ether by comparing the velocities of light from different directions. In an attempt to overcome this difficulty he introduced in 1895 the idea of local time (different time rates in different locations). Lorentz arrived at the notion that moving bodies approaching the velocity of light contract in the direction of motion. The Irish physicist George Francis FitzGerald had already arrived at this notion independently, and in 1904 Lorentz extended his work and developed the Lorentz transformations. These mathematical formulas describe the increase of mass, shortening of length, and dilation of time that are characteristic of a moving body and form the basis of Einstein's special theory of relativity. In 1912 Lorentz became director of research at the Teyler Institute, Haarlem, though he remained honorary professor at Leiden, where he gave weekly lectures.

ERNST MACH

(b. Feb. 18, 1838, Chirlitz-Turas, Moravia, Austrian Empire—d. Feb. 19, 1916, Haar, Ger.)

Austrian physicist and philosopher Ernst Mach established important principles of optics, mechanics, and wave dynamics and supported the view that all knowledge is a conceptual organization of the data of sensory experience (or observation).

Mach was educated at home until the age of 14, then went briefly to gymnasium (high school) before entering the University of Vienna at 17. He received his doctorate in physics in 1860 and taught mechanics and physics in Vienna until 1864, when he became professor of mathematics at the University of Graz. Mach's interests had already begun to turn to the psychology and physiology of sensation, although he continued to identify himself as a physicist and to conduct physical research throughout his career. During the 1860s he discovered the physiological phenomenon that has come to be called Mach's bands, the tendency of the human eye to see bright or dark bands near the boundaries between areas of sharply differing illumination.

Mach left Graz to become professor of experimental physics at the Charles University in Prague in 1867, remaining there for the next 28 years. There he conducted studies on kinesthetic sensation, the feeling associated with movement and acceleration. Between 1873 and 1893 he developed optical and photographic techniques for the measurement of sound waves and wave propagation. In 1887 he established the principles of supersonics and the Mach number—the ratio of the velocity of an object to the velocity of sound.

In *Beiträge zur Analyse der Empfindungen* (1886; *Contributions to the Analysis of the Sensations,* 1897), Mach advanced the concept that all knowledge is derived from sensation; thus, phenomena under scientific investigation can be understood only in terms of experiences, or "sensations," present in the observation of the phenomena. This view leads to the position that no statement in natural science is admissible unless it is empirically verifiable. Mach's exceptionally rigorous criteria of verifiability led him to reject such metaphysical concepts as absolute time and space, and prepared the way for the Einstein relativity theory.

Mach also proposed the physical principle, known as Mach's principle, that inertia (the tendency of a body at rest to remain at rest and of a body in motion to continue in motion in the same direction) results from a relationship of that object with all the rest of the matter in the universe. Inertia, Mach argued, applies only as a function of the interaction between one body and other bodies in the universe, even at enormous distances. Mach's inertial theories also were cited by Einstein as one of the inspirations for his theories of relativity.

Mach returned to the University of Vienna as professor of inductive philosophy in 1895, but he suffered a stroke two years later and retired from active research in 1901, when he was appointed to the Austrian parliament. He continued to lecture and write in retirement, publishing *Erkenntnis und Irrtum* ("Knowledge and Error") in 1905 and an autobiography in 1910.

A.A. MICHELSON

(b. Dec. 19, 1852, Strelno, Prussia [now Strzelno, Pol.]—d. May 9, 1931, Pasadena, Calif., U.S.)

German-born American physicist Albert Abraham Michelson established the speed of light as a fundamental constant and pursued other spectroscopic and metrological investigations. He received the 1907 Nobel Prize for Physics.

Michelson came to the United States with his parents when he was two years old. From New York City, the family made its way to Virginia City, Nev., and San Francisco, where the elder Michelson prospered as a merchant. At 17, Michelson entered the United States Naval Academy at Annapolis, Md., where he did well in science but was rather below average in seamanship. He graduated in 1873, then served as science instructor at the academy from 1875 until 1879.

In 1878 Michelson began work on what was to be the passion of his life, the accurate measurement of the speed of light. He was able to obtain useful values with home-made apparatuses. Feeling the need to study optics before he could be qualified to make real progress, he traveled to Europe in 1880 and spent two years in Berlin, Heidelberg, and Paris, resigning from the U.S. Navy in 1881. Upon his return to the United States, he determined the velocity of light to be 299,853 km (186,329 miles) per second, a value that remained the best for a generation, until Michelson bettered it.

While in Europe, Michelson began constructing an interferometer, a device designed to split a beam of light in two, send the parts along perpendicular paths, then bring them back together. If the light waves had, in the interim, fallen out of step, interference fringes of alternating light and dark bands would be obtained. From the width and number of those fringes, unprecedentedly delicate measurements could be made, comparing the velocity of light rays traveling at right angles to each other.

It was Michelson's intention to use the interferometer to measure Earth's velocity against the "ether" that was then thought to make up the basic substratum of the universe. If Earth were traveling through the light-conducting ether, then the speed of the light traveling in the same direction would be expected to be equal to the velocity of light plus the velocity of Earth, whereas the speed of light traveling at right angles to Earth's path would be expected to travel only at the velocity of light. His earliest experiments in Berlin showed no interference fringes, however, which seemed to signify that there was no difference in the speed of the light rays and, therefore, no Earth motion relative to the ether.

In 1883 he accepted a position as professor of physics at the Case School of Applied Science in Cleveland and there concentrated his efforts on improving the delicacy of his interferometer experiment. By 1887, with the help of his colleague, American chemist Edward Williams Morley, he was ready to announce the results of what has since come to be called the Michelson-Morley experiment. Those results were still negative; there were no interference fringes and apparently no motion of Earth relative to the ether.

It was perhaps the most significant negative experiment in the history of science. In terms of classical Newtonian physics, the results were paradoxical. Evidently, the speed of light plus any other added velocity was still equal only to the speed of light. To explain the result of the Michelson-Morley experiment, physics had to be recast on a new and more refined foundation, something that resulted, eventually, in Albert Einstein's formulation of the theory of relativity in 1905.

In 1892 Michelson, after serving as professor of physics at Clark University at Worcester, Mass., from 1889, was appointed professor and the first head of the department of physics at the newly organized University of Chicago, a position he held until his retirement in 1929. From 1923 to 1927 he served as president of the National Academy of Sciences. In 1907 he became the first American ever to receive a Nobel Prize in the sciences, for his spectroscopic and metrological investigations, the first of many honours he was to receive.

Michelson advocated using some particular wavelength of light as a standard of distance (a suggestion generally accepted in 1960) and, in 1893, measured the standard metre in terms of the red light emitted by heated cadmium. His interferometer made it possible for him to determine the width of heavenly objects by matching the light rays from the two sides and noting the interference fringes that resulted. In 1920, using a 6-metre (20-foot) interferometer attached to a 254-centimetre (100-inch) telescope, he succeeded in measuring the diameter of the star Betelgeuse (Alpha Orionis) as 386,160,000 km (300 times the diameter of the Sun). This was the first substantially accurate determination of the size of a star.

In 1923 Michelson returned to the problem of the accurate measurement of the velocity of light. In the California mountains he surveyed a 35-km pathway between two mountain peaks, determining the distance to an accuracy of less than 2.5 cm. He made use of a special eight-sided revolving mirror and obtained a value of 299,798 km/sec for the velocity of light. To refine matters further, he made use of a long, evacuated tube through which a light beam was reflected back and forth until it had traveled 16 km through a vacuum. Michelson died before the results of his final tests could be evaluated, but in 1933 the final figure was announced as 299,774 km/sec, a value less than 2 km/sec higher than the value accepted in the 1970s.

HERMANN MINKOWSKI

(b. June 22, 1864, Aleksotas, Russian Empire [now in Kaunas, Lithuania] — d. Jan. 12, 1909, Göttingen, Ger.)

German mathematician Hermann Minkowski developed the geometrical theory of numbers and made numerous contributions to number theory, mathematical physics, and the theory of relativity. His idea of combining the three dimensions of physical space with that of time into a four-dimensional "Minkowski space" — space-time — laid the mathematical foundations for Albert Einstein's special theory of relativity.

The son of German parents living in Russia, Minkowski returned to Germany with them in 1872 and spent his youth in the royal Prussian city of Königsberg. A gifted prodigy, he began his studies at the University of Königsberg and the University of Berlin at age 15. Three years later he was awarded the "Grand Prix des Sciences Mathématiques" by the French Academy of Sciences for his paper on the representation of numbers as a sum of five squares. During his teenage years in Königsberg he met and befriended another young mathematical prodigy, David Hilbert, with whom he worked closely both at Königsberg and later at the University of Göttingen.

After earning his doctorate in 1885, Minkowski taught mathematics at the Universities of Bonn (1885–94), Königsberg (1894–96), Zürich (1896–1902), and Göttingen (1902–09). Together with Hilbert, he pursued research on the electron theory of the Dutch physicist Hendrik Lorentz and its modification in Einstein's special theory of relativity. In *Raum und Zeit* (1907; "Space and Time") Minkowski gave his famous four-dimensional geometry based on the group of Lorentz transformations of special relativity theory. His major work in number theory was *Geometrie der Zahlen* (1896; "Geometry of Numbers"). His works were collected in David Hilbert (ed.), *Gesammelte Abhandlungen*, 2 vol. (1911; "Collected Papers").

EDWARD WILLIAMS MORLEY

(b. Jan. 29, 1838, Newark, N.J., U.S.—d. Feb. 24, 1923, West Hartford, Conn.)

American chemist Edward Williams Morley is best known for his collaboration with the physicist A.A. Michelson in an attempt to measure the relative motion of Earth through a hypothetical ether.

Morley graduated from Williams College in 1860 and then pursued both scientific and theological studies. He took up a Congregational pastorate in Ohio in 1868 and in the following year joined the faculty of Western Reserve College, remaining with the school when it moved to Cleveland in 1882 and became Western Reserve University. He continued to teach there until his retirement in 1906. From 1873 to 1888 he also taught at the Cleveland Medical School.

Morley's personal research centred on questions requiring precise determinations of the density and atomic weight of various gases, especially of oxygen. His reputation as a skilled experimenter attracted the attention of Michelson, then at the nearby Case School of Applied Science. In 1887 the pair performed what have come to be known as the Michelson-Morley experiments, which failed definitively to detect any "ether-drag" effect on the speed of light measured in various directions relative to the motion of Earth. This result was a major step leading toward Albert Einstein's special theory of relativity.

WOLFGANG PAULI

(b. April 25, 1900, Vienna, Austria—d. Dec. 15, 1958, Zürich, Switz.)

Austrian-born physicist Wolfgang Ernst Friedrich Pauli won the 1945 Nobel Prize for Physics for his discovery in 1925 of the Pauli exclusion principle, which states that in an atom no two electrons can occupy the same quantum state simultaneously. Pauli made major contributions to quantum mechanics, quantum field theory, and solid-state physics, and he successfully hypothesized the existence of the neutrino.

In addition to his original work, he wrote masterful syntheses of several areas of physical theory that are considered classics of scientific literature. An even deeper influence was left by his personal interactions with other scientists, as recorded by numerous testimonies and a vast but never dull extant correspondence.

Pauli was raised among the intellectual elite of Vienna, a highly cosmopolitan city that was one of the most important centres of scientific advancement at the turn of the 20th century. Pauli's godfather and mentor was the physicist-philosopher Ernst Mach, for whom he was given one of his middle names. Pauli later wrote that Mach's influence in his upbringing was an "anti-metaphysical baptism."

Having demonstrated outstanding mathematical abilities—Pauli taught himself the then new theory of relativity in his gymnasium years and published his first paper on the subject when he was 18—he enrolled in physics at the University of Munich, where he studied the most advanced physics of the day: the Bohr-Sommerfeld quantum theory of the atom, under Arnold Sommerfeld. Pauli distinguished himself not only for his brilliance but also for his exacting rigour and impertinent witticisms. A review of the theory of relativity that he wrote for *Encyklopädie der mathematischen Wissenschaften* ("Encyclopedia of Mathematical Sciences") in 1921 gained him early fame and high praise from Albert Einstein.

After completing a doctorate in theoretical physics in 1921, Pauli worked as an assistant to Max Born at the University of Göttingen (1921–22) and as an assistant to Wilhelm Lenz at the University of Hamburg (1922). Pauli took a one-year leave to work at Niels Bohr's Institute for Theoretical Physics (1922–23) in Copenhagen before returning to Hamburg in 1924 to complete his habilitation (a postdoctoral degree that is required in order to hold a professorship in most European universities).

With Pauli's return to Hamburg in 1924 as a lecturer, he participated in the creation of quantum mechanics. He first solved (1924–25) certain vexing difficulties in the theory of atomic spectra by the introduction of a new quantum number—a quantity that was later called spin but that Pauli, in accordance with his philosophical rejection of visualizable models, called "a two-valuedness not describable classically." He concluded that if a quantum state so defined was occupied by one electron, it was excluded for other electrons. This rule was eventually incorporated in the quantum mechanics of multiparticle systems and elevated to a principle, the exclusion principle. It thus became the foundation of the Fermi-Dirac statistics, the branch of quantum statistics developed by Italian physicist Enrico Fermi and English physicist P.A.M. Dirac, of quantum chemistry, and of the quantum theory of solids. In addition to this and other results in quantum mechanics, quantum field theory, and the theory of magnetism, Pauli assisted the work of others with his critical input. (He later came to be called the "conscience of physics" for his critical insights.) His analysis of the philosophical foundations and methodology of physics played a central role in the so-called Copenhagen interpretation of quantum mechanics, based on the renunciation of causality and on the affirmation of the positivist notion that physical concepts must be limited by the possibilities of observation.

In 1928 Pauli obtained a professorship at the Swiss Federal Institute of Technology (Eidgenössische Technische Hochschule, or ETH) in Zürich, a position that he kept for the rest of his life and from which, together with German physicist Gregor Wentzel of the University of Zürich, he created a successful "school" of theoretical physics. The suicide of Pauli's mother in 1927 and the end of his marriage with the Berlin dancer Käthe Deppner in 1930 led him to psychoanalysis, first as a therapy (an analyst was recommended to him by the Swiss psychologist Carl Gustav Jung) and then as a philosophical interest. He studied Jung's ideas and their import on the scientific understanding of the physical world, which led him to distance himself from positivism. Although Pauli ended two years of personal therapy with Jung in 1934, when Pauli married Franciska (Franca) Bertram, the two men developed an extensive correspondence through the following years concerning physics and psychology.

In 1930 Pauli conjectured the existence of neutral particles (later called neutrinos) to preserve the conservation of energy in nuclear beta decay. (Experimental detection of the neutrino did not come until 1956.) His analysis of symmetries in quantum field theory resulted in the formulation in 1940 of the spin-statistics theorem, which established a necessary connection between the spin of a particle and its statistical properties.

The outbreak of World War II and the possible threat of Nazi persecution (Pauli's paternal grandparents were Jews) led him to accept a visiting professorship in 1940 at the Institute for Advanced Study in Princeton, N.J., U.S., where he worked mainly on meson theory. Although he became an American citizen in 1946, he went back to Europe that year, first to finally accept his 1945 Nobel Prize and then to return to his former position at ETH in Zürich. Back at ETH he worked on renormalization in quantum electrodynamics, in collaboration with his students, and on the CPT (charge, parity, time) symmetry in quantum field theory. He finally became a Swiss citizen in 1949.

In his later years Pauli's involvement with philosophy intensified, while travels and epistolary exchanges kept alive his dialogue with Einstein, Bohr, and others. A rich correspondence with Jung, as well as several publications, testify to Pauli's ongoing quest to understand "physis and psyche as complementary aspects of the same reality."

MAX PLANCK

(b. April 23, 1858, Kiel, Schleswig [Germany]—d. Oct. 4, 1947, Göttingen, W.Ger.)

Theoretical physicist Max Karl Ernst Ludwig Planck originated quantum theory, which won him the Nobel Prize for Physics in 1918.

Planck made many contributions to theoretical physics, but his fame rests primarily on his role as originator of the quantum theory. This theory revolutionized our understanding of atomic and subatomic processes, just as Albert Einstein's theory of relativity revolutionized our understanding of space and time. Together they constitute the fundamental theories of 20th-century physics. Both have forced man to revise some of his most cherished philosophical beliefs, and both have led to industrial and military applications that affect every aspect of modern life.

Max Planck.
Encyclopædia Britannica, Inc.

Planck was the sixth child of a distinguished jurist and professor of law at the University of Kiel. The long family tradition of devotion to church and state, excellence in scholarship, incorruptibility, conservatism, idealism, reliability, and generosity became deeply ingrained in Planck's own life and work. When Planck was nine years old, his father received an appointment at the University of Munich, and Planck entered the city's renowned Maximilian Gymnasium, where a teacher, Hermann Müller, stimulated his interest in physics and mathematics. But Planck excelled in all subjects, and after graduation at age 17 he faced a difficult career decision. He ultimately chose physics over classical philology or music because he had dispassionately reached the conclusion that it was in physics that his greatest originality lay. Music, nonetheless, remained an integral part of his life. He possessed the gift of absolute pitch and was an excellent pianist who daily found serenity and delight at the keyboard, enjoying especially the works of Schubert and Brahms. He also loved the outdoors, taking long walks each day and hiking and climbing in the mountains on vacations, even in advanced old age.

Planck entered the University of Munich in the fall of 1874 but found little encouragement there from physics professor Philipp von Jolly. During a year spent at the University of Berlin (1877–78), he was unimpressed by the lectures of Hermann von Helmholtz and Gustav Robert Kirchhoff, despite their eminence as research scientists. His intellectual capacities were, however, brought to a focus as the result of his independent study, especially of Rudolf Clausius' writings on thermodynamics. Returning to Munich, he received his doctoral degree in July 1879 (the year of Einstein's birth) at the unusually young age of 21. The following year he completed his *Habilitationsschrift* (qualifying dissertation) at Munich and became a *Privatdozent* (lecturer). In 1885, with the help of his father's professional connections, he was appointed *ausserordentlicher Professor* (associate professor) at the University of Kiel. In 1889, after the death of Kirchhoff, Planck received an appointment to the University of Berlin, where he came to venerate Helmholtz as a mentor and colleague. In 1892 he was promoted to *ordentlicher Professor* (full professor). He had only nine doctoral students altogether, but his Berlin lectures on all branches of theoretical physics went through many editions and exerted great influence. He remained in Berlin for the rest of his active life.

Planck recalled that his "original decision to devote myself to science was a direct result of the discovery... that the laws of human reasoning coincide with the laws governing the sequences of the impressions we receive from the world about us; that, therefore, pure reasoning can enable man to gain an insight into the mechanism of the [world]...." He deliberately decided, in other words, to become a theoretical physicist at a time when theoretical physics was not yet recognized as a discipline in its own right. But he went further: he concluded that the existence of physical laws presupposes that the "outside world is something independent from man, something absolute, and the quest for the laws which apply to this absolute appeared...as the most sublime scientific pursuit in life."

The first instance of an absolute in nature that impressed Planck deeply, even as a Gymnasium student, was the law of the conservation of energy, the first law of thermodynamics. Later, during his university years, he became equally convinced that the entropy law, the second law of thermodynamics, was also an absolute law of nature. The second law became the subject of his doctoral dissertation at Munich, and it lay at the core of the researches that led him to discover the quantum of action, now known as Planck's constant h, in 1900.

In 1859–60 Kirchhoff had defined a blackbody as an object that reemits all of the radiant energy incident upon it; i.e., it is a perfect emitter and absorber of radiation. There was, therefore, something absolute about blackbody radiation, and by the 1890s various experimental and theoretical attempts had been made to determine its spectral energy distribution—the curve displaying how much radiant energy is emitted at different frequencies for a given temperature of the blackbody. Planck was particularly attracted to the formula found in 1896 by his colleague Wilhelm Wien at the Physikalisch-Technische Reichsanstalt (PTR) in Berlin-Charlottenburg, and he subsequently made a series of attempts to derive "Wien's law" on the basis of the second law of thermodynamics. By October 1900, however, other colleagues at the PTR, the experimentalists Otto Richard Lummer, Ernst Pringsheim, Heinrich Rubens, and Ferdinand Kurlbaum, had found definite indications that Wien's law, while valid at high frequencies, broke down completely at low frequencies.

Planck learned of these results just before a meeting of the German Physical Society on October 19. He knew how the entropy of the radiation had to depend mathematically upon its energy in the high-frequency region if Wien's law held there. He also saw what this dependence had to be in the low-frequency region in order to reproduce the experimental results there. Planck guessed, therefore, that he should try to combine these two expressions in the simplest way possible, and to transform the result into a formula relating the energy of the radiation to its frequency.

The result, which is known as Planck's radiation law, was hailed as indisputably correct. To Planck, however, it was simply a guess, a "lucky intuition." If it was to be taken seriously, it had to be derived somehow from first principles. That was the task to which Planck immediately directed his energies, and by Dec. 14, 1900, he had succeeded—but at great cost. To achieve his goal, Planck found that he had to relinquish one of his own most cherished beliefs, that the second law of thermodynamics was an absolute law of nature. Instead he had to embrace Ludwig Boltzmann's interpretation, that the second law was a statistical law. In addition, Planck had to assume that the oscillators comprising the blackbody and re-emitting the radiant energy incident upon them could not absorb this energy continuously but only in discrete amounts, in quanta of energy; only by statistically distributing these quanta, each containing an amount of energy $h\nu$ proportional to its frequency, over all of the oscillators present in the blackbody could Planck derive the formula he had hit upon two months earlier. He adduced additional evidence for the importance of his formula by using it to evaluate the constant h (his value was 6.55×10^{-34} joule-second, close to the modern value), as well as the so-called Boltzmann constant (the fundamental constant in kinetic theory and statistical mechanics), Avogadro's number, and the charge of the electron. As time went on physicists recognized ever more clearly that—because Planck's constant was not zero but had a small but finite value—the microphysical world, the world of atomic dimensions, could not in principle be described by ordinary classical mechanics. A profound revolution in physical theory was in the making.

Planck's concept of energy quanta, in other words, conflicted fundamentally with all past physical theory. He was driven to introduce it strictly by the force of his logic; he was, as one historian put it, a reluctant revolutionary. Indeed, it was years before the far-reaching consequences of Planck's achievement were generally recognized, and in this Einstein played a central role. In 1905, independently of Planck's work, Einstein argued that under certain circumstances radiant energy itself seemed to consist of quanta (light quanta, later called photons), and in 1907 he showed the generality of the quantum hypothesis by using it to interpret the temperature dependence of the specific heats of solids. In 1909 Einstein introduced the wave–particle duality into physics. In October 1911 he was among the group of prominent physicists who attended the first Solvay conference in Brussels. The discussions there stimulated Henri Poincaré to provide a mathematical proof that Planck's radiation law necessarily required the introduction of quanta—a proof that converted James (later Sir James) Jeans and others into supporters of the quantum theory. In 1913 Niels Bohr also contributed greatly to its establishment through his quantum theory of the hydrogen atom. Ironically, Planck himself was one of the last to struggle for a return to classical theory, a stance he later regarded not with regret but as a means by which he had thoroughly convinced himself of the necessity of the quantum theory. Opposition to Einstein's radical light quantum hypothesis of 1905 persisted until after the discovery of the Compton effect in 1922.

Planck was 42 years old in 1900 when he made the famous discovery that in 1918 won him the Nobel Prize for Physics and that brought him many other honours. It is not surprising that he subsequently made no discoveries of comparable importance. Nevertheless, he continued to contribute at a high level to various branches of optics, thermodynamics and statistical mechanics, physical chemistry, and other fields. He was also the first prominent physicist to champion Einstein's special theory of relativity (1905). "The velocity of light is to the Theory of Relativity," Planck remarked, "as the elementary quantum of action is to the Quantum Theory; it is its absolute core."

In 1914 Planck and the physical chemist Walther Hermann Nernst succeeded in bringing Einstein to Berlin, and after the war, in 1919, arrangements were made for Max von Laue, Planck's favourite student, to come to Berlin as well. When Planck retired in 1928, another prominent theoretical physicist, Erwin Schrödinger, the originator of wave mechanics, was chosen as his successor. For a time, therefore, Berlin shone brilliantly as a centre of theoretical physics—until darkness enveloped it in January 1933 with the ascent of Adolf Hitler to power.

In his later years, Planck devoted more and more of his writings to philosophical, aesthetic, and religious questions. Together with Einstein and Schrödinger, he remained adamantly opposed to the indeterministic, statistical worldview introduced by Bohr, Max Born, Werner Heisenberg, and others into physics after the advent of quantum mechanics in 1925–26. Such a view was not in harmony with Planck's deepest intuitions and beliefs. The physical universe, Planck argued, is an objective entity existing independently of man; the observer and the observed are not intimately coupled, as Bohr and his school would have it.

Planck became permanent secretary of the mathematics and physics sections of the Prussian Academy of Sciences in 1912 and held that position until 1938; he was also president of the Kaiser Wilhelm Society (now the Max Planck Society) from 1930 to 1937. These offices and others placed Planck in a position of great authority, especially among German physicists; seldom were his decisions or advice questioned. His authority, however, stemmed fundamentally not from the official appointments he held but from his personal moral force. His fairness, integrity, and wisdom were beyond question. It was completely in character that Planck went directly to Hitler in an attempt to reverse Hitler's devastating racial policies and that he chose to remain in Germany during the Nazi period to try to preserve what he could of German physics.

Planck was a man of indomitable will. Had he been less stoic, and had he had less philosophical and religious conviction, he could scarcely have withstood the tragedies that entered his life after age 50. In 1909, his first wife, Marie Merck, the daughter of a Munich banker, died after 22 years of happy marriage, leaving Planck with two sons and twin daughters. The elder son, Karl, was killed in action in 1916. The following year, Margarete, one of his daughters, died in childbirth, and in 1919 the same fate befell Emma, his other daughter. World War II brought

Henri Poincaré, 1909. H. Roger-Viollet

further tragedy. Planck's house in Berlin was completely destroyed by bombs in 1944. Far worse, the younger son, Erwin, was implicated in the attempt made on Hitler's life on July 20, 1944, and in early 1945 he died a horrible death at the hands of the Gestapo. That merciless act destroyed Planck's will to live. At war's end, American officers took Planck and his second wife, Marga von Hoesslin, whom he had married in 1910 and by whom he had had one son, to Göttingen. There, in 1947, in his 89th year, he died. Death, in the words of James Franck, came to him "as a redemption."

HENRI POINCARÉ

(b. April 29, 1854, Nancy, France—d. July 17, 1912, Paris)

French mathematician Jules Henri Poincaré was one of the greatest mathematicians and mathematical physicists at the end of 19th century. He made a series of profound innovations in geometry, the theory of differential equations, electromagnetism, topology, and the philosophy of mathematics.

Poincaré grew up in Nancy and studied mathematics from 1873 to 1875 at the École Polytechnique in Paris. He continued his studies at the Mining School in Caen before receiving his doctorate from the École Polytechnique in 1879. While a student, he discovered new types of complex functions that solved a wide variety of differential equations. This major work involved one of the first "mainstream" applications of non-Euclidean geometry, a subject discovered by the Hungarian János Bolyai and the Russian Nikolay Lobachevsky about 1830 but not generally accepted by mathematicians until the 1860s and '70s. Poincaré published a long series of papers on this work in 1880–84 that effectively made his name internationally.

The prominent German mathematician Felix Klein, only five years his senior, was already working in the area, and it was widely agreed that Poincaré came out the better from the comparison.

In the 1880s Poincaré also began work on curves defined by a particular type of differential equation, in which he was the first to consider the global nature of the solution curves and their possible singular points (points where the differential equation is not properly defined). He investigated such questions as: Do the solutions spiral into or away from a point? Do they, like the hyperbola, at first approach a point and then swing past and recede from it? Do some solutions form closed loops? If so, do nearby curves spiral toward or away from these closed loops? He showed that the number and types of singular points are determined purely by the topological nature of the surface. In particular, it is only on the torus that the differential equations he was considering have no singular points.

Poincaré intended this preliminary work to lead to the study of the more complicated differential equations that describe the motion of the solar system. In 1885 an added inducement to take the next step presented itself when King Oscar II of Sweden offered a prize for anyone who could establish the stability of the solar system. This would require showing that equations of motion for the planets could be solved and the orbits of the planets shown to be curves that stay in a bounded region of space for all time. Some of the greatest mathematicians since Isaac Newton had attempted to solve this problem, and Poincaré soon realized that he could not make any headway unless he concentrated on a simpler, special case, in which two massive bodies orbit one another in circles around their common centre of gravity while a minute third body orbits them both. The third body is taken to be so small that it does not affect the orbits of the larger ones. Poincaré could establish that the orbit is stable, in the sense that the small body returns infinitely often arbitrarily close to any position it has occupied. This does not mean, however, that it does not also move very far away at times, which would have disastrous consequences for life on Earth. For this and other achievements in his essay, Poincaré was awarded the prize in 1889. But, on writing the essay for publication, Poincaré discovered that another result in it was wrong, and in putting that right he discovered that the motion could be chaotic. He had hoped to show that if the small body could be started off in such a way that it traveled in a closed orbit, then starting it off in almost the same way would result in an orbit that at least stayed close to the original orbit. Instead, he discovered that even small changes in the initial conditions could produce large, unpredictable changes in the resulting orbit. (This phenomenon is now known as pathological sensitivity to initial positions, and it is one of the characteristic signs of a chaotic system.) Poincaré summarized his new mathematical methods in astronomy in *Les Méthodes nouvelles de la mécanique céleste*, 3 vol. (1892, 1893, 1899; "The New Methods of Celestial Mechanics").

Poincaré was led by this work to contemplate mathematical spaces (now called manifolds) in which the position of a point is determined by several coordinates. Very little was known about such manifolds, and, although the German mathematician Bernhard Riemann had hinted at them a generation or more earlier, few had taken the hint. Poincaré took up the task and looked for ways in which such manifolds could be distinguished, thus opening up the whole subject of topology, then known as analysis situs. Riemann had shown that in two dimensions surfaces can be distinguished by their genus (the number of holes in the surface), and Enrico Betti in Italy and Walther von Dyck in Germany had extended this work to three dimensions, but much remained to be done. Poincaré singled out the idea of considering closed curves in the manifold that cannot be deformed into one another. For example, any curve on the surface of a sphere can be continuously shrunk to a point, but there are curves on a torus (curves wrapped around a hole, for instance) that cannot. Poincaré asked if a three-dimensional manifold in which every curve can be shrunk to a point is topologically equivalent to a three-dimensional sphere. This problem (now known as the Poincaré conjecture) became one of the most important unsolved problems in algebraic topology. Ironically, the conjecture was first proved for dimensions greater than three: in dimensions five and above by Stephen Smale in the 1960s and in dimension four as a consequence of work by Simon Donaldson and Michael Freedman in the 1980s. Finally, Grigori Perelman proved the conjecture for three dimensions in 2006. All of these achievements were marked with the award of a Fields Medal. Poincaré's *Analysis Situs* (1895) was an early systematic treatment of topology, and he is often called the father of algebraic topology.

Poincaré's main achievement in mathematical physics was his magisterial treatment of the electromagnetic theories of Hermann von Helmholtz, Heinrich Hertz, and Hendrik Lorentz. His interest in this topic—which, he showed, seemed to contradict Newton's laws of mechanics—led him to write a paper in 1905 on the motion of the electron. This paper, and others of his at this time, came close to anticipating Albert Einstein's discovery of the theory of special relativity. But Poincaré never took the decisive step of reformulating traditional concepts of space and time into space-time, which was Einstein's most profound achievement. Attempts were made to obtain a Nobel Prize in physics for Poincaré, but his work was too theoretical and insufficiently experimental for some tastes.

About 1900 Poincaré acquired the habit of writing up accounts of his work in the form of essays and lectures for the general public. Published as *La Science et l'hypothèse* (1903; *Science and Hypothesis*), *La Valeur de la science* (1905; *The Value of Science*), and *Science et méthode* (1908; *Science and Method*), these essays form the core of his reputation as a philosopher of mathematics and science. His most famous claim in this connection is that much of science is a matter of convention. He came to this view on thinking about the nature of space: Was it Euclidean or non-Euclidean? He argued that one could never tell, because one could not logically separate the physics involved from the mathematics, so any choice would be a matter of convention. Poincaré suggested that one would naturally choose to work with the easier hypothesis.

Poincaré's philosophy was thoroughly influenced by psychologism. He was always interested in what the human mind understands, rather than what it can formalize. Thus, although Poincaré recognized that Euclidean and non-Euclidean geometry are equally "true," he argued that our experiences have and will continue to predispose us to formulate physics in terms of Euclidean geometry; Einstein proved him wrong. Poincaré also felt that our understanding of the natural numbers was innate and therefore fundamental, so he was critical of attempts to reduce all of mathematics to symbolic logic (as advocated by Bertrand Russell in England and Louis Couturat in France) and of attempts to reduce mathematics to axiomatic set theory. In these beliefs he turned out to be right, as shown by Kurt Gödel in 1931.

In many ways Poincaré's influence was extraordinary. All the topics discussed above led to the creation of new branches of mathematics that are still highly active today, and he also contributed a large number of more technical results. Yet in other ways his influence was slight. He never attracted a group of students around him, and the younger generation of French mathematicians that came along tended to keep him at a respectful distance. His failure to appreciate Einstein helped to relegate his work in physics to obscurity after the revolutions of special and general relativity. His often imprecise mathematical exposition, masked by a delightful prose style, was alien to the generation in the 1930s who modernized French mathematics under the collective pseudonym of Nicolas Bourbaki, and they proved to be a powerful force. His philosophy of mathematics lacked the technical aspect and profundity of developments inspired by the German mathematician David Hilbert's work. However, its diversity and fecundity has begun to prove attractive again in a world that sets more store by applicable mathematics and less by systematic theory.

Most of Poincaré's original papers are published in the 11 volumes of his *Oeuvres de Henri Poincaré* (1916–54). In 1992 the Archives–Centre d'Études et de Recherche Henri-Poincaré founded at the University of Nancy 2 began to edit Poincaré's scientific correspondence, signaling a resurgence of interest in him.

ERWIN SCHRÖDINGER

(b. Aug. 12, 1887, Vienna, Austria—d. Jan. 4, 1961, Vienna)

Austrian theoretical physicist Erwin Schrödinger contributed to the wave theory of matter and to other fundamentals of quantum mechanics. He shared the 1933 Nobel Prize for Physics with the British physicist P.A.M. Dirac.

Schrödinger entered the University of Vienna in 1906 and obtained his doctorate in 1910, upon which he accepted a research post at the university's Second Physics Institute. He saw military service in World War I and then went to the University of Zürich in 1921, where he remained for the next six years. There, in a six-month period in 1926, at the age of 39, a remarkably late age for original work by theoretical physicists, he produced the papers that gave the foundations of quantum wave mechanics. In those papers he described his partial differential equation that is the basic equation of quantum mechanics and bears the same relation to the mechanics of the atom as Newton's equations of motion bear to planetary astronomy. Adopting a proposal made by Louis de Broglie in 1924 that particles of matter have a dual nature and in some situations act like waves, Schrödinger introduced a theory describing the behaviour of such a system by a wave equation that is now known as the Schrödinger equation. The solutions to Schrödinger's equation, unlike the solutions to Newton's equations, are wave functions that can only be related to the probable occurrence of physical events. The definite and readily visualized sequence of events of the planetary orbits of Newton is, in quantum mechanics, replaced by the more abstract notion of probability. (This aspect of the quantum theory made Schrödinger and several other physicists profoundly unhappy, and he devoted much of his later life to formulating philosophical objections to the generally accepted interpretation of the theory that he had done so much to create.)

In 1927 Schrödinger accepted an invitation to succeed Max Planck, the inventor of the quantum hypothesis, at the University of Berlin, and he joined an extremely distinguished faculty that included Albert Einstein. He remained at the university until 1933, at which time he reached the decision that he could no longer live in a country in which the persecution of Jews had become a national policy. He then began a seven-year odyssey that took him to Austria, Great Britain, Belgium, the Pontifical Academy of Science in Rome, and—finally in 1940—the Dublin Institute for Advanced Studies, founded under the influence of Premier Eamon de Valera, who had been a mathematician before turning to politics. Schrödinger remained in Ireland for the next 15 years, doing research both in physics and in the philosophy and history of science. During this period he wrote *What Is Life?* (1944), an attempt to show how quantum physics can be used to explain the stability of genetic structure. Although much of what Schrödinger had to say in this book has been modified and amplified by later developments in molecular biology, his book remains one of the most useful and profound introductions to the subject. In 1956 Schrödinger retired and returned to Vienna as professor emeritus at the university.

Of all of the physicists of his generation, Schrödinger stands out because of his extraordinary intellectual versatility. He was at home in the philosophy and literature of all of the Western languages, and his popular scientific writing in English, which he had learned as a child, is among the best of its kind. His study of ancient Greek science and philosophy, summarized in his *Nature and the Greeks* (1954), gave him both an admiration for the Greek invention of the scientific view of the world and a skepticism toward the relevance of science as a unique tool with which to unravel the ultimate mysteries of human existence. Schrödinger's own metaphysical outlook, as expressed in his last book, *Meine Weltansicht* (1961; *My View of the World*), closely paralleled the mysticism of the Vedānta.

Because of his exceptional gifts, Schrödinger was able in the course of his life to make significant contributions to nearly all branches of science and philosophy, an almost unique accomplishment at a time when the trend was toward increasing technical specialization in these disciplines.

KARL SCHWARZSCHILD

(b. Oct. 9, 1873, Frankfurt am Main, Ger.—d. May 11, 1916, Potsdam)

Karl Schwarzschild was a German astronomer whose contributions, both practical and theoretical, were of primary importance in the development of 20th-century astronomy.

Schwarzschild's exceptional ability in science became evident at the age of 16, when his paper on the theory of celestial orbits was published. In 1901 he became professor and director of the observatory at the University of Göttingen, and eight years later he was appointed director of the Astrophysical Observatory at Potsdam.

While at Göttingen, Schwarzschild introduced precise methods in photographic photometry. The results of his studies clearly demonstrated the relationship between the spectral type and colour of a star. He pioneered in the use of a coarse grating (for example, a glass plate with closely spaced parallel lines etched into it) in the course of measurement of the separation of double stars; the technique has found widespread use in determining stellar magnitude and colour. He also developed certain basic methods for the analysis of solar spectra obtained during eclipses.

Schwarzschild enunciated the principle of radiative equilibrium and was the first to recognize clearly the role of radiative processes in the transport of heat in stellar atmospheres. His hypothesis of stellar motion is one of the most important results to come out of his fundamental work in modern statistical methods in astronomy. He also made theoretical studies of the pressure exerted on small, solid particles by radiation.

Schwarzschild made fundamental contributions to theoretical physics and to relativity. He was one of the great pioneers in developing the theory of atomic spectra proposed by Niels Bohr. Independently of Arnold Sommerfeld, Schwarzschild developed the general rules of quantization, gave the complete theory of the Stark effect (the effect of an electric field on light), and initiated the quantum theory of molecular spectra.

Schwarzschild gave the first exact solution of Albert Einstein's general gravitational equations, which led to a description of the geometry of space in the neighbourhood of a mass point. He also laid the foundation of the theory of black holes by using the general equations to demonstrate that bodies of sufficient mass would have an escape velocity exceeding the speed of light and, therefore, would not be directly observable.

While serving in the imperial German army during World War I, Schwarzschild contracted a fatal illness.

JULIAN SEYMOUR SCHWINGER

(b. Feb. 12, 1918, New York, N.Y., U.S.—d. July 16, 1994, Los Angeles, Calif.)

American physicist Julian Seymour Schwinger was the joint winner, with Richard P. Feynman and Tomonaga Shin'ichirō, of the Nobel Prize for Physics in 1965 for introducing new ideas and methods into quantum electrodynamics.

Schwinger was a child prodigy, publishing his first physics paper at age 16. He earned a bachelor's degree (1937) and a doctorate (1939) from Columbia University in New York City, before engaging in postdoctoral studies at the University of

Albert Einstein (left) *presenting the first Albert Einstein Award for achievement in natural sciences to Austrian mathematician Kurt Gödel* (second from right) *and American physicist Julian Schwinger* (right), *with Lewis L. Stauss looking on, March 14, 1951.* New York World-Telegram and the Sun Newspaper/Library of Congress, Washington, D.C. (Digital ID cph 3c33518)

California at Berkeley with physicist J. Robert Oppenheimer. Schwinger left Berkeley in the summer of 1941 to accept an instructorship at Purdue University, West Lafayette, Ind., and in 1943 he joined the Radiation Laboratory at the Massachusetts Institute of Technology, where many scientists had been assembled to help with wartime research on radar. In the fall of 1945 Schwinger accepted an appointment at Harvard University and in 1947 became one of the youngest full professors in the school's history. From 1972 until his death, Schwinger was a professor in the physics department at the University of California at Los Angeles.

Schwinger was one of the participants at the meeting held in June 1947 on Shelter Island, Long Island, N.Y., at which reliable experimental data were presented that contradicted the predictions of the English theoretical physicist P.A.M. Dirac's relativistic quantum theory of the electron. In particular, experimental data contradicted Dirac's prediction that certain hydrogen electron stationary states were degenerate (i.e., had the same energy as certain other states) as well as Dirac's prediction for the value of the magnetic moment of the electron. Schwinger made a quantum electrodynamical calculation that made use of the notions of mass and charge renormalization, which brought agreement between theory and experimental data. This was a crucial breakthrough that initiated a new era in quantum field theory. Richard Feynman and Tomonaga Shin'ichirō independently had carried out similar calculations, and in 1965 the three of them shared the Nobel Prize. Their work created a new and very successful quantum mechanical description of the interaction between electrically charged entities and the electromagnetic field that conformed with the principles of Albert Einstein's special theory of relativity.

Schwinger's work extended to almost every frontier of modern theoretical physics. He had a profound influence on physics both directly and through being the academic adviser for more than 70 doctoral students and more than 20 postdoctoral fellows, many of whom became the outstanding theorists of their generation.

ARNOLD SOMMERFELD

(b. Dec. 5, 1868, Königsberg, Prussia [now Kaliningrad, Russia]—d. April 26, 1951, Munich, Ger.)

Arnold Johannes Wilhelm Sommerfeld was a German physicist whose atomic model permitted the explanation of fine-structure spectral lines.

After studying mathematics and science at Königsberg University, Sommerfeld became an assistant at the University of Göttingen and then taught mathematics at Clausthal (1897) and Aachen (1900).

As professor of theoretical physics at Munich (1906–31), he did his most important work. His investigations of atomic spectra led him to suggest that, in the Bohr model of the atom, the electrons move in elliptical orbits as well as circular ones. From this idea he postulated the azimuthal quantum number. He later introduced the magnetic quantum number as well. Sommerfeld also did detailed work on wave mechanics, and his theory of electrons in metals proved valuable in the study of thermoelectricity and metallic conduction.

OTTO STERN

(b. Feb. 17, 1888, Sohrau, Ger. [now Zory, Pol.]—d. Aug. 17, 1969, Berkeley, Calif., U.S.)

German-born scientist Otto Stern was a winner of the Nobel Prize for Physics in 1943 for his development of the molecular beam as a tool for studying the characteristics of molecules and for his measurement of the magnetic moment of the proton.

Stern's early scientific work was theoretical studies of statistical thermodynamics. In 1914 he became a lecturer in theoretical physics at the University of Frankfurt and in 1923 a professor of physical chemistry at the University of Hamburg. Stern and Walther Gerlach performed their historic molecular-beam experiment at Hamburg in the early 1920s. By shooting a beam of silver atoms through a nonuniform magnetic field onto a glass plate, they found that the beam split into two distinct beams instead of broadening into a continuous band. This experiment

Otto Stern at the presentation of the Nobel Prizes, New York City, 1943. Encyclopædia Britannica, Inc.

verified the space quantization theory, which stated that atoms can align themselves in a magnetic field only in a few directions (two for silver), instead of in any direction, as classical physics had suggested. In 1933 Stern measured the magnetic moment (strength of a subatomic particle's magnetic property) of the proton by using a molecular beam and found that it was actually about 2 ½ times the theoretical value.

In 1933, when the Nazis rose to power, Stern was compelled to leave Germany. He went to the United States, where he became research professor of physics at the Carnegie Institute of Technology, Pittsburgh. He remained there until his retirement in 1945.

TOMONAGA SHIN'ICHIRŌ

(b. March 31, 1906, Kyōto, Japan—d. July 8, 1979, Tokyo)

Japanese physicist Tomonaga Shin'ichirō was a joint winner, with Richard P. Feynman and Julian S. Schwinger of the United States, of the Nobel Prize for Physics in 1965 for developing basic principles of quantum electrodynamics.

Tomonaga became professor of physics at Bunrika University (later Tokyo University of Education) in 1941, the year he began his investigations of the problems of quantum electrodynamics. World War II isolated him from Western scientists, but in 1943 he completed and published his research. Tomonaga's theoretical work made quantum electrodynamics (the theory of the interactions of charged subatomic particles with the electromagnetic field) consistent with the theory of special relativity. It was only after the war, in 1947, that his work came to the attention of the West, at about the same time that Feynman and Schwinger published the results of their research.

It was found that all three had achieved essentially the same result from different approaches and had resolved the inconsistencies of the old theory without making any drastic changes.

Tomonaga was president of the Tokyo University of Education from 1956 to 1962, and the following year he was named chairman of the Japan Science Council. Throughout his life Tomonaga actively campaigned against the spread of nuclear weapons and urged that resources be spent on the peaceful use of nuclear energy. Most notable of his works available in English translation are *Quantum Mechanics* (1962) and his Nobel lecture *Development of Quantum Electrodynamics: Personal Recollections* (1966).

GEORGE EUGENE UHLENBECK

(b. Dec. 6, 1900, Batavia, Java [now Jakarta, Indon.] — d. Oct. 31, 1988, Boulder, Colo., U.S.)

Dutch American physicist George Eugene Uhlenbeck, with Samuel A. Goudsmit, proposed the concept of electron spin.

In 1925, while working on his Ph.D. at the University of Leiden, Neth. (1927), he and Goudsmit put forth their idea of electron spin after ascertaining that electrons rotate about an axis. Uhlenbeck joined the physics department at the University of Michigan, U.S., in 1927, returned to the Netherlands, as professor at the State University at Utrecht, and then became full professor at the University of Michigan in 1939. From 1943 to 1945 he worked at the Radiation Laboratory of the Massachusetts Institute of Technology, and in the postwar period he worked in the Netherlands. In 1960 he was appointed professor and physicist at the Rockefeller Medical Research Center at the State University of New York, New York City, becoming professor emeritus in 1974. He wrote many papers on atomic structure, quantum mechanics, kinetic theory of matter, and nuclear physics.

WILHELM WIEN

(b. Jan. 13, 1864, Gaffken, Prussia [now Parusnoye, Russia]—d. Aug. 30, 1928, Munich, Ger.)

German physicist Wilhelm Carl Werner Otto Fritz Franz Wien received the Nobel Prize for Physics in 1911 for his displacement law concerning the radiation emitted by the perfectly efficient blackbody (a surface that absorbs all radiant energy falling on it).

Wien obtained his doctorate at the University of Berlin in 1886 and soon began to work on the problem of radiation. Although the radiation emitted from a blackbody is distributed over a wide range of wavelengths, there is an intermediate wavelength at which the radiation reaches a maximum. In 1893 Wien stated in his law that this maximum wavelength is inversely proportional to the absolute temperature of the body. Because the accuracy of Wien's law declined for longer wavelengths, Max Planck was led to further investigations culminating in his quantum theory of radiation.

Wien was appointed professor of physics at the University of Giessen in 1899 and at the University of Munich in 1920. He also made contributions in the study of cathode rays (electron beams), X-rays, and canal rays (positively charged atomic beams). His autobiography was published under the title *Aus dem Leben und Wirken eines Physikers* (1930; "From the Life and Work of a Physicist").

CONCLUSION

In classical physics, space is conceived as having the absolute character of an empty stage in which events in nature unfold as time flows onward independently; events occurring simultaneously for one observer are presumed to be simultaneous for any other; mass is taken as impossible to create or destroy; and a particle given sufficient energy acquires a velocity that can increase without limit. The special and general theory of relativity, developed principally by Albert Einstein and now so adequately confirmed by experiment as to have the status of physical law, shows that all these, as well as other apparently obvious assumptions, are false.

Special and general relativity have profoundly affected physical science and human existence, most dramatically in applications of nuclear energy and nuclear weapons. Additionally, relativity and its rethinking of the fundamental categories of space and time have provided a basis for certain philosophical, social, and artistic interpretations that have influenced human culture in different ways.

The behaviour of matter and radiation on the atomic scale often seems peculiar, and the consequences of quantum theory are accordingly difficult to understand and to believe. Its concepts frequently conflict with common-sense notions derived from observations of the everyday world. There is no reason, however, why the behaviour of the atomic world should conform to that of the familiar, large-scale world. It is important to realize that quantum mechanics is a branch of physics and that the business of physics is to describe and account for the way the world—on both the large and the small scale—actually is and not how one imagines it or would like it to be.

The study of quantum mechanics is rewarding for several reasons. First, it illustrates the essential methodology of physics. Second, it has been enormously successful in giving correct results in practically every situation to which it has been applied. There is, however, an intriguing paradox. In spite of the overwhelming practical success of quantum mechanics, the foundations of the subject contain unresolved problems—in particular, problems concerning the nature of measurement. An essential feature of quantum mechanics is that it is generally impossible, even in principle, to measure a system without disturbing it; the detailed nature of this disturbance and the exact point at which it occurs are obscure and controversial. Thus, quantum mechanics has attracted some of the ablest scientists, and they have erected what is perhaps the finest intellectual edifice of the period.

Glossary

astrophysics Using physical and chemical knowledge to understand the nature of celestial objects and their physical processes.

binary star A pair of stars in orbit around a common centre of gravity.

black hole A region of space in which the gravitational field is so powerful that nothing, including light, can escape its pull.

calculus Branch of mathematics, developed by Isaac Newton and Gottfried Wilhelm Leibniz, concerned with the calculation of instantaneous rates of change and the summation of infinitely many small factors to determine some whole.

cosmological constant Term reluctantly added by Albert Einstein to his equations of general relativity in order to obtain a solution to the equations that described a static universe, as he believed it to be at the time.

cosmology The field of study that brings together the natural sciences, particularly astronomy and physics, in a joint effort to understand the physical universe as a unified whole.

dark matter A component of the universe whose presence is discerned from its gravitational attraction rather than its luminosity.

empirical A term that denotes information gained by means of observation, experience, or experiment.

Euclidean geometry The study of plane and solid figures on the basis of axioms and theorems employed by the Greek mathematician Euclid.

force A push or pull that can cause an object with mass to change its velocity.

gravity well The field of gravitational potential around a massive body.

inertia Property of a body by virtue of which it opposes any agency that attempts to put it in motion or, if it is moving, to change the magnitude or direction of its velocity.

integer Any positive or negative whole number or zero.

isotropic Identical in direction and with regard to physical properties.

kinetic energy The extra energy that an object gains while in motion.

light-year The distance that light waves travel in one Earth year.

luminosity Amount of light emitted by an object in a unit of time.

lunar eclipse An eclipse that occurs whenever the Moon passes behind the earth such that Earth blocks the Sun's rays from striking the Moon.

neutrino Elementary subatomic particle with no electric charge, very little mass, and 1/2 unit of spin.

photons Packets of electromagnetic radiation.

pulsar Neutron stars that emit pulses of radiation once per rotation.

quantum A discrete natural unit, or packet, of energy, charge, angular momentum, or other physical property.

quantum theory The study of reactions between matter and radiation.

redshift The increase in the wavelength of electromagnetic radiation received by a detector compared with the wavelength emitted by a receding source.

supernova A violently exploding star whose luminosity after eruption suddenly increases to many times its normal level.

superstring theory A theory that attempts to merge quantum mechanics with Albert Einstein's general theory of relativity.

theorem A proposition or statement that is demonstrated; a statement to be proved.

velocity Quantity that defines how fast and in what direction an object is moving.

wavelength Distance between corresponding points of two consecutive waves.

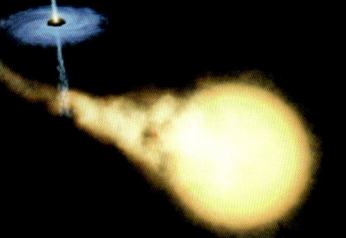

BIBLIOGRAPHY

Albert Einstein, *Relativity: The Special and General Theory*, trans. from the German by Robert W. Lawson (1916, reissued 2001), is a concise presentation with some mathematics. Martin Gardner, *Relativity Simply Explained* (1962, reissued 1997), is more expansive and less mathematical.

Works for readers with physics background at the college level include Edwin F. Taylor and John Archibald Wheeler, *Spacetime Physics: Introduction to Special Relativity*, 2nd ed. (1992, reissued 1997), and *Exploring Black Holes: Introduction to General Relativity* (1995, reissued 2001); and Ray A. d'Inverno, *Introducing Einstein's Relativity* (1997).

The philosophical meaning of relativity is presented in Hans Reichenbach, *Philosophy of Space and Time*, trans. by Maria Reichenbach and John Freund (1958; originally published in German, 1928); and Lawrence Sklar, *Space, Time, and Spacetime* (1974, reissued 1977). The historical context for relativity is discussed in Helge Kragh, *Quantum Generations: A History of Physics in the Twentieth Century* (1999, reissued 2002).

An outstanding work containing an account of the special theory of relativity is Abraham Pais, *"Subtle Is the Lord—": The Science and Life of Albert Einstein* (1982). Some good introductions at the undergraduate level are W. Rindler, *Essential Relativity: Special, General, and Cosmological*, 2nd ed. (1977); and James H. Smith, *Introduction to Special Relativity* (1965). More substantial treatises are J. Aharoni, *The Special Theory of Relativity*,

2nd ed. (1965, reprinted 1985); and J.L. Synge, *Relativity: The Special Theory*, 2nd ed. (1965).

Several book-length studies have been written on the historical development of quantum mechanics; especially noteworthy are Olivier Darrigol, *From C-Numbers to Q-Numbers: The Classical Analogy in the History of Quantum Theory* (1992); and Max Jammer, *The Conceptual Development of Quantum Mechanics*, 2nd ed. (1989).

Careful historical and philosophical studies of the work of many of the early architects of quantum theory may be found in Thomas S. Kuhn, *Black-Body Theory and the Quantum Discontinuity, 1894–1912* (1978, reprinted 1987); Bruce R. Wheaton, *The Tiger and the Shark: Empirical Roots of Wave-Particle Dualism* (1983, reissued 1991); Abraham Pais, *Niels Bohr's Times: In Physics, Philosophy, and Polity* (1991); Arthur Fine, *The Shaky Game: Einstein, Realism, and the Quantum Theory*, 2nd ed. (1996); Max Dresden, *H.A. Kramers: Between Tradition and Revolution* (1987); David C. Cassidy, *Uncertainty: The Life and Science of Werner Heisenberg* (1992); Walter Moore, *Schrödinger: Life and Thought* (1989); and Dugald Murdoch, *Niels Bohr's Philosophy of Physics* (1987, reissued 1990). The birth of quantum theory in the period 1900–26, primarily within German university circles, is nicely contextualized by Christa Jungnickel and Russell McCormmach, *Intellectual Mastery of Nature: Theoretical Physics from Ohm to Einstein*, 2 vol. (1986, reissued 1990). The transition from nonrelativistic quantum mechanics to renormalized quantum electrodynamics over the period 1926–49 is traced by Silvan S. Schweber, *QED and the Men Who Made It: Dyson, Feynman, Schwinger, and Tomonaga* (1994).

There are a number of excellent texts on quantum mechanics at the undergraduate and graduate level. The following is a selection, beginning with the more elementary: A.P. French and Edwin F. Taylor, *An Introduction to Quantum Physics* (1978); Alastair I.M. Rae, *Quantum Mechanics*, 2nd ed. (1986); Richard L. Liboff, *Introductory Quantum Mechanics*, 2nd ed. (1992); Eugen Merzbacher, *Quantum Mechanics*, 2nd ed. (1970); J.J. Sakurai, *Modern Quantum Mechanics*, rev. ed. (1994); and Anthony Sudbery, *Quantum Mechanics and the Particles of Nature: An Outline for Mathematicians* (1986), rather mathematical but including useful accounts and summaries of quantum metaphysics. Richard P. Feynman, Robert B. Leighton, and Matthew Sands, *The Feynman Lectures on Physics*, vol. 3, *Quantum Mechanics* (1965), is a personal and stimulating look at the subject. A good introduction to quantum electrodynamics is Richard P. Feynman, *QED: The Strange Theory of Light and Matter* (1985).

J.C. Polkinghorne, *The Quantum World* (1984); John Gribbin, *In Search of Schrödinger's Cat: Quantum Physics and Reality* (1984); Heinz R. Pagels, *The Cosmic Code: Quantum Physics as the Language of Nature* (1982); and David Z. Albert, *Quantum Mechanics and Experience* (1992), are all highly readable and instructive books written at a popular level. Bernard d'Espagnat, *Conceptual Foundations of Quantum Mechanics*, 2nd ed. (1976), is a technical account of the fundamental conceptual problems involved. The proceedings of a conference, *New Techniques and Ideas in Quantum Measurement Theory*, ed. by Daniel M. Greenberger (1986), contain a wide-ranging set of papers that deal with both the experimental and theoretical aspects of the measurement problem.

Applications are presented by H. Haken and H.C. Wolf, *Atomic and Quantum Physics: An Introduction to the Fundamentals of Experiment and Theory*, 2nd enlarged ed. (1987; originally published in German, 2nd rev. and enlarged ed., 1983); Emilio Segrè, *Nuclei and Particles: An*

Introduction to Nuclear and Subnuclear Physics, 2nd rev. and enlarged ed. (1977, reissued 1980); Donald H. Perkins, *Introduction to High Energy Physics*, 3rd ed. (1987); Charles Kittel, *Introduction to Solid State Physics*, 6th ed. (1986); and Rodney Loudon, *The Quantum Theory of Light*, 2nd ed. (1983). B.W. Petley, *The Fundamental Physical Constants and the Frontier of Measurement* (1985), gives a good account of the fundamental constants.

INDEX

A

Abbott, Edwin, 48
ABC of Relativity, 50
Adenauer, Konrad, 189, 190
Aharonov-Bohm effect, 119
Alpher, Ralph, 175
Alsos, 180
Alsos, 180
Amaldi, Edoardo, 164
American Association for the Advancement of Science, 136
American Physical Society, 136
Analysis Situs, 218
Anderson, Carl David, 69, 112–113, 140
Aristotle, 1, 38, 50
Aspect, Alain, 98
Astrophysical Observatory, 223
Atkinson, R., 174
Atomic Energy Commission (AEC), 136, 168
Atomic Energy States, 180
Auger, Pierre-Victor, 76
Auger effect, 75–76

B

Bacher, Robert F., 180
Balasz, Margit, 142
Bell, John Stewart, 97, 119
Bell's inequality, 97–98, 119–120
Bell Telephone Laboratories, 137, 180
Ben-Gurion, David, 160
Bernstein, Aaron, 147
Bertram, Franciska, 206
Besso, Michele, 148, 149, 152
Bethe, Hans, 87–88, 113–118, 170, 175
Betti, Enrico, 217
big bang, 29, 36, 40, 43, 44, 174, 175
blackbodies, 52–53, 210, 211, 231
Blackett, Patrick M.S., 112
black holes, 29, 34–35, 49, 50, 160, 163, 224
Bohm, David, 93, 95, 118–120
Bohr, Niels, 54–59, 60, 63, 75, 88, 94, 119, 120–128, 139, 160, 183, 184, 185, 188, 204, 207, 212, 213, 224, 227
Boltzmann, Ludwig, 211
Bolyai, János, 215
Bondi, Hermann, 38
Born, Max, 60, 63, 128–132, 139, 154, 163, 183, 185, 190, 191, 204, 213
Bose, Satyendra Nath, 70, 109–111, 132, 163, 164
Bothe, Walther, 128, 176
Bourbaki, Nicolas, 220
Bragg, William Lawrence, 115

British Royal Society, 135
Broglie, Louis-Victor de, 59–60, 61, 62, 84, 91, 93, 132–135, 137, 179, 180, 221
Broken Scale Variance and the Light Cone, 178
Brookhaven National Laboratory, 180
Brown, Gerald, 118

C

California Institute of Technology (Caltech), 112, 113, 118, 171, 172, 177
Carnegie Institute of Technology, 229
Case School of Applied Science, 200, 203
Causality and Chance in Modern Physics, 119
cesium clock, 104–107
Chadwick, James, 126, 165
Chaplin, Charlie, 155–156
Character of Physical Law, The, 173
Chu, Steven, 110
Churchill, Winston, 127
Claiborne, Robert, 180
Clausius, Rudolf, 208
Cohen-Tannoudji, Claude, 110
Compton, Arthur Holly, 58–59, 86
Compton effect, 86, 176, 195, 212
Condon, Edward, 77, 135–137
Consciousness and the Physical World, 194
Contributions to the Analysis of the Sensations, 197
Copenhagen Institute of Theoretical Physics, 174
Copenhagen interpretation, 94, 99–100, 124, 185, 205
Copernicus, 37
Cornell, Eric, 109, 110
Coster, Dirk, 123
Coulomb's constant, 66
Coulomb's law, 31–32, 55
Couturat, Louis, 219
Craven, Thomas, 48
Creation of the Universe, The, 176
Crommelin, Andrew, 30
Cubism, 48–49
curved space-time and geometric gravitation, 26–28

D

Danish Committee for the Support of Refugee Intellectual Workers, 126
Davisson, Clinton Joseph, 59, 134, 137, 179, 180
Deppner, Käthe, 206
Development of Quantum Electrodynamics, 230
Dicke, Robert, 192
Dirac equation, 64, 69, 140
Dirac, Paul A.M., 64, 69, 78, 86, 87, 88, 110, 130, 137–142, 164, 184, 185, 191, 205, 220, 226
Donaldson, Simon, 218
Dublin Institute for Advanced Studies, 222
Dyck, Walther von, 218
Dynamics of Crystal Lattices, 130

E

Eddington, Sir Arthur Stanley, 29–30, 49, 143–146

Ehrenberg, Hedwig, 130
Ehrenfest, Paul, 153, 157, 163
Eightfold Way, 178
Einstein, Albert, 1, 4–5, 6, 7, 8, 9,
   10, 11, 12, 23, 24–25, 26, 27, 28,
   30, 32, 33, 37, 38, 39, 40, 43–45,
   46, 47, 48, 49, 50, 53–54, 93,
   94–95, 96, 98, 109–111, 119,
   124, 130, 132, 133, 134, 138, 144,
   146–163, 166, 167, 173, 185,
   195, 196, 198, 200, 202, 203,
   204, 207, 208, 212, 213, 218,
   219, 220, 221, 224, 226, 232
Einstein, Eduard, 150, 157
Einstein, Hermann, 146, 148
Einstein-de Sitter model, 43–45
Einstein-Hilbert action, 153
Eisenhower, Dwight D., 117
electron spin and antiparticles,
   64–69
Emergency Association for
   German Science, 189
Emergency Committee of
   Atomic Scientists, 159
Ending of Time, The, 120
Enrico Fermi Award, 163
EPR thought experiment, 94–95,
   96, 160–161
equivalence, principle of, 24–26
Esaki, Leo, 192, 193
Euclid, 43–44, 50, 144, 153, 215, 219
European Council for Nuclear
   Research, (CERN), 128, 189
Everett, Hugh, III, 100
Expanding Universe, The, 144

F

Federal Bureau of Investigation
   (FBI), 119, 158

Fermi, Alberto, 163
Fermi, Enrico, 70, 110, 115, 130,
   139, 163–169, 205
Fermi, Laura, 166
Fermi-Dirac statistics, 110, 139,
   164, 205
Fermilab, 163
Feynman diagrams, 170, 171, 172
*Feynman Lectures on Physics,
   The*, 172
Feynman, Richard, 16, 17, 22, 88,
   142, 169–173, 224, 226, 229
Fields Medal, 218
FitzGerald, George Francis, 196
Flatland, 48
four-dimensional space-time,
   12–22
Fowler, Ralph, 115, 138, 139
Franck, James, 57, 130, 135, 190, 215
Franck-Condon principle, 135
Freedman, Michael, 218
French Academy of Sciences,
   135, 202
French Atomic Energy
   Commissariat, 135
Fresnel, Augustin-Jean, 52
Freud, Sigmund, 156
Friedmann, Aleksandr A., 28,
   39–40, 41, 43, 44, 49,
   173–174, 175
Frisch, Otto Robert, 125, 158, 166
"From the Life and Work of a
   Physicist," 231
Fundamental Theory, 145

G

Galileo, 1, 6, 7, 9
Gamow, George, 77, 174–176
Gedankenexperiments, 4–5, 25

Geiger, Hans, 55, 176–177
Geiger-Müller counter, 165, 176, 177
Gell-Mann, Murray, 102, 103, 168, 171, 177–179
General Advisory Committee (GAC), 168
"Geometry of Numbers," 202
Gerlach, Walther, 64, 65, 81, 100, 179, 227
German National Institute for Science and Technology, 176
German Physical Society, 210
German Research Council, 189
Germer, Lester Halbert, 59, 134, 137, 179–180
Giaever, Ivar, 192, 193
Gödel, Kurt, 219
Goeppert-Mayer, Maria, 71, 130
Gold, Thomas, 38
Goudsmit, Samuel A., 64, 65–66, 180, 230
Grossmann, Marcel, 149
Gurney, Ronald W., 77

H

Hahn, Otto, 125, 158, 165
Hartree, Douglas R., 74
Hau, Lene, 111
Heisenberg, August, 182
Heisenberg uncertainty principle, 83–87, 182, 184
Heisenberg, Werner, 60, 61, 83–87, 115, 123–124, 126, 130, 131, 139, 182–190, 191, 213
Heitler, Walter, 130
Helmholtz, Hermann von, 208, 209, 218

Henri Poincaré Institute, 135
Hertz, Gustav, 57
Hertz, Heinrich, 218
Hess, Victor Francis, 112
Hevesy, Georg, 123, 125
hidden variables, 92–94, 96, 97, 119
Hilbert, David, 153, 202, 220
Himmler, Heinrich, 186
Hitler, Adolf, 115, 126, 192, 213, 215
Hoesslin, Marga von, 215
Hoover, J. Edgar, 119, 158
House Un-American Activities Committee (HUAC), 119, 136
Houtermans, F., 174
Hoyle, Fred, 38
Hubble, Edwin, 29, 38, 40, 42, 44, 45, 155, 175
Hubble's constant, 44, 45
Hulse, Russell, 33
Humboldt Foundation, 189–190

I

identical particles and multi-electron atoms, 69–74
Institute for Advanced Study, 157, 206
Institute for Nuclear Studies, 177
Institute of Theoretical Physics, 122–123, 183, 184, 185, 195, 204
Internal Constitution of the Stars, 146
International Astronomical Union, 145
International Business Machines Corporation (IBM), 193
"Investigation of the State of Aether in Magnetic Fields, The," 147

J

Japan Science Council, 230
Jeans, Sir James, 145, 212
Jensen, J. Hans D., 71
JILA, 109, 111
Jin, Deborah, 111
Joliet-Curie, Frédéric, 165
Joliet-Curie, Irène, 165
Jolly, Philipp von, 208
Jordan, Pascual, 60, 130, 131, 139, 183, 185, 190–192
Josephson, Brian D., 107, 192–194
Josephson effect, 107–108, 192, 193
Jung, Carl, 206, 207

K

Kaiser Wilhelm Institute for Physics (KWI), 152, 187, 188, 189, 213
Kalinga Prize, 135
Kant, Immanuel, 50
Kármán, Theodore von, 131
Kepler, Johannes, 2
Ketterle, Wolfgang, 109, 111
Kirchhoff, Gustav Robert, 208, 209, 210
Klein, Felix, 130, 216
Koch, Pauline, 146
Krogh, August, 125
Kurlbaum, Ferdinand, 210

L

Lamb, Willis E., Jr., 88
Landau, Lev D., 171
LaPorte, Paul, 49
Larmor, Joseph, 130
Laser Interferometer Gravitational-Wave Observatory (LIGO), 34
Laser Interferometer Space Antenna (LISA), 34
Laue, Max von, 58, 194–195, 213
Lawrence, Ernest O., 125
Lemaître, Georges, 39–40, 144, 175
Lenard, Philipp, 157
Lenz, Wilhelm, 204
"Let's Call It Plectics," 178
Lobachevsky, Nikolay, 215
London, Fritz, 130
Lorentz, Hendrik, 9, 68, 151, 195–196, 202, 218
Lorentz transformations, 9, 13, 15, 16–17, 196, 202
Los Alamos Laboratory, 116, 118, 127, 136, 158, 167–168, 169, 170
Löwenthal, Elsa, 152, 157
Lummer, Otto Richard, 210

M

Mach, Ernst, 4, 196–198, 204
Mach's bands, 197
Majorana, Ettore, 164
Manhattan Project, 116, 118, 158, 167–168, 170, 187, 188
Maric, Mileva, 149, 150, 151, 152
Marsden, Ernest, 55
Massachusetts Institute of Technology (MIT), 109, 116, 170, 177, 180, 225, 230
Mathematical Association, 145
Mathematical Theory of Relativity, The, 144
Max Planck Institute, 189, 190, 195, 213

Maxwell, James Clerk, 3, 5, 32, 87, 150, 152, 170, 195
Mayakovsky, Vladimir, 49
McCarthyism, 119
Meitner, Lise, 125, 158, 166
Merck, Marie, 214
meson, decay of a, 101–104
Michelson, A.A., 3, 4, 6, 50, 198–201, 202–203
Michelson-Morley experiment, 4, 6, 50, 196, 200, 203
Millikan, Robert Andrews, 112
Milne, Edward A., 37, 145
Minkowski, Hermann, 12, 13, 16, 17, 130, 201–202
Minkowski space, 201–202
Modern Theme, The, 48
Morley, Edward, 3, 4, 6, 50, 196, 200, 202–203
Mossbauer effect, 193
Mount Wilson Observatory, 155
Mr. Tomkins in Wonderland, 176
Müller, Hermann, 208
Müller, Walther, 177
Mussolini, Benito, 164
My View of the World, 222

N

National Academy of Sciences, 200
National Accelerator Laboratory, 163
National Bureau of Standards, 136
National Institute of Standards and Technology (NIST), 109
Nature and the Greeks, 222
Nature of the Physical World, The, 144

Nazism, 126, 156–157, 166, 185, 186, 187, 188, 192, 206, 214, 229
Nernst, Walther Hermann, 213
"New Methods of Celestial Mechanics, The," 217
New Pathways of Science, 144
Newton, Isaac, 1–2, 6, 7, 8, 17, 19, 24, 25, 26, 29, 31, 32, 37, 38, 45, 48, 50, 51, 56, 62, 140, 142, 150, 152, 153, 155, 200, 216, 218, 221
Nobel Prize, 33, 47, 68, 109, 110, 112, 113, 116, 120, 122, 123, 128, 131, 132, 135, 137, 138, 140, 146, 152, 154, 155, 157, 161, 163, 166, 169, 177, 180, 182, 185, 186, 192, 194, 195, 196, 198, 200, 203, 206, 207, 212, 218, 220, 224, 226, 227, 229, 230, 231
Nordic Institute for Atomic Physics, 128
Nørlund, Margrethe, 122
Nuclear Test Ban Treaty, 117

O

observables, incompatible, 80–83
Occhialini, Giuseppe, 112
One Hundred Authors Against Einstein, 157
One, Two, Three...Infinity, 176
"On Quantum Mechanics," 183–184
"On the Perceptual Content of Quantum Theoretical Kinematics and Mechanics," 184

Oppenheimer, J. Robert, 118, 127, 130, 136, 159–160, 167, 225
Optics, 50
Order of Merit, 145
"Origin of Chemical Elements, The," 175
Ortega y Gasset, José, 48
Ossietzky, Carl von, 186

P

Pais, Abraham, 102, 103
Pauli exclusion principle, 70–71, 110, 164, 203
Pauli, Wolfgang, 70, 71, 110, 115, 130, 139, 164, 165, 185, 191, 203–207
Pauling, Linus, 180
Pavlovsk Aerological Observatory, 173
Perelman, Grigori, 218
Phillips, William D., 110
Philosophy of Physical Science, The, 144
Physical Society, 144, 145
Picasso, Pablo, 48
Planck, Max, 46, 47, 53, 54, 66, 108, 130, 133, 141, 152, 184, 189, 190, 195, 207–215, 221, 231
Planck's constant, 53, 66, 141, 184, 210
Planck's Law and the Hypothesis of Light Quanta, 132
Planet Called Earth, A, 176
Podolsky, Boris, 94–95, 160
Poincaré, Henri, 4, 135, 151, 212, 215–220
Pontecorvo, Bruno, 164
Pontifical Academy of Science, 222

Popular Books on Physical Science, 147
President's Science Advisory Committee (PSAC), 117
Principles of Quantum Mechanics, The, 78, 141
Pringsheim, Ernst, 210
Prussian Academy of Sciences, 213

Q

QED: The Strange Theory of Light and Matter, 173
Quantum Electrodynamics, 172
Quantum Mechanics, 230
"Quantum-Theoretical Reinterpretation of Kinematic and Mechanical Relations," 183
Quantum Theory, 119
quantum voltage standard, 107–109
Quark and the Jaguar, The, 178–179

R

Rabi, Isidor, 168, 180
Ramachandran, V.S., 194
Rasetti, Franco, 164
Reagan, Ronald, 117
relativity, overview of, 1–23
Relativity Theory of Protons and Electrons, 144–145
Report on the Relativity Theory of Gravitation, 144
Retherford, Robert, 88
Reviews of Modern Physics, 115
Riemann, Bernhard, 217

Road from Los Alamos, The, 117
Rockefeller Foundation, 115, 125, 163, 191
Rockefeller Medical Research Center, 231
Roosevelt, Franklin D., 127, 158, 167
Rosen, Nathan, 94–95, 160
Royal Astronomical Society, 145, 154
Royal Greenwich Observatory, 30, 143
Royal Society, 154, 194
Rubens, Heinrich, 210
Russell, Bertrand, 50, 219
Rutherford, Ernest, 55, 56, 122, 176
Rydberg constant, 55, 57

S

Sachs, Alexander, 158
Santa Fe Institute, 178
Schrödinger, Erwin, 60, 61–64, 73, 74–76, 78, 80, 86, 91, 98–99, 100, 131, 134, 138, 139, 184, 185, 191, 213, 220–223
Schrödinger equation, 62, 64, 73, 74–76, 78, 80, 91, 98, 99, 100, 131, 139, 184, 221
Schumacher, Elisabeth, 187
Schwarzschild, Karl, 29, 223–224
Schwinger, Julian, 87, 169, 224–226, 229
Science and Hypothesis, 219
Science and Method, 219
Science and the Unseen World, 144
Scientific Study of Unidentified Flying Objects, The, 137
Segrè, Emilio, 164
Sitter, Willem de, 39, 43–45, 173

Smale, Stephen, 218
Solvay Conferences, 124, 152, 212
Sommerfeld, Arnold, 60, 115, 182, 183, 186, 204, 224, 226–227
Space, Time and Gravitation, 144
Speer, Albert, 188
Spinoza, Benedict de, 156
Stanford Linear Accelerator, 172
Star Called the Sun, A, 176
Stark effect, 68–69, 224
Stark, Johannes, 68–69, 157, 186
Stars and Atoms, 146
Stellar Movements and the Structure of the Universe, 143–144
Stern, Otto, 64, 65, 81, 100, 179, 227–229
Stern-Gerlach experiment, 64–65, 81, 100, 179
Strassmann, Fritz, 125, 158, 165
Structure of Line Spectra, The, 180
Stückelberg, Ernest C.G., 16, 17, 22
Swiss Federal Institute of Technology (ETH), 205, 206
Szilard, Leo, 158, 167

T

Tagore, Rabindranath, 156
Talmud, Max, 147
Taylor, Joseph H., Jr., 33
Teller, Edward, 116, 117, 175
Teller-Ulam mechanism, 117
Teyler Institute, 196
Theory of Fundamental Processes, The, 172
Thomson, George, 134, 137
Thomson, J.J., 130, 145, 154

Time, 180
Tomonaga Shin'ichirō, 88, 169, 224, 226, 229–230
Truman, Harry S., 168–169
tunneling, 76–78, 193
twin paradox, 4, 11–12

U

Uhlenbeck, George E., 64, 65–66, 180, 230–231
universe, expansion of, 28–29, 38, 41, 45, 144–145, 155, 173–174, 175, 192
Uranium Committee, 158

V

Value of Science, The, 219

W

Washington Conferences on Theoretical Physics, 116
Weber, Heinrich, 149
Wecklein, Anna, 182
Weisskopf, Victor, 130
Wentzel, Gregor, 205
What Is Life?, 222
Wheeler, John, 28, 126, 170
Whittaker, Sir Edmund Taylor, 145
Wholeness and the Implicate Order, 120
Wieman, Carl, 109, 110–111
Wien, Wilhelm, 53, 210, 231
Wigner, Eugene, 142, 177, 191
Wilson, K., 178
Winteler, Jost, 148
wormholes, 35, 49, 160

Y

Young, Thomas, 51, 90

Z

Zeeman effect, 67–68, 69, 183, 196
Zeeman, Pieter, 68, 195, 196